SIMPLY
SCIENCE

DK | Penguin Random House

Produced for DK by
cobalt id
www.cobaltid.co.uk

Design Director Paul Reid
Designers Clare Joyce, Phil Gamble, Mik Gates
Editor Marek Walisiewicz
Creative Technical Support Darren Bland

Senior Editor Peter Frances
Senior Designer Jessica Tapolcai
Managing Editor Angeles Gavira Guerrero
Managing Art Editor Michael Duffy
Production Editor Andy Hilliard
Senior Production Controller Meskerem Berhane
Jacket Design Development Manager
Sophia M.T.T
Project Jacket Designer Juhi Sheth
Publishing Director Liz Wheeler
Art Director Maxine Pedliham
Managing Director Liz Gough
Design Director Phil Ormerod

First published in Great Britain in 2024 by
Dorling Kindersley Limited
DK, One Embassy Gardens, 8 Viaduct Gardens,
London SW11 7BW

The authorised representative in the EEA is
Dorling Kindersley Verlag GmbH. Arnulfstr. 124,
80636 Munich, Germany

Copyright © 2024 Dorling Kindersley Limited
A Penguin Random House Company
10 9 8 7 6 5 4 3 2 1
001–336903–Nov/2024

A CIP catalogue record for this book
is available from the British Library.
ISBN: 978-0-2416-3475-2

Printed and bound in Slovakia

www.dk.com

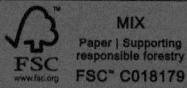

MIX
Paper | Supporting
responsible forestry
FSC™ C018179

This book was made with Forest
Stewardship Council™ certified
paper – one small step in DK's
commitment to a sustainable future.
Learn more at www.dk.com/uk/
information/sustainability

CONTENTS

CHEMISTRY

PHYSICS

BIOLOGY

EARTH

ASTRONOMY

CONSULTANTS

Andy Extance is a science writer. His work explores everything related to chemistry, from Earth's environment to space, from food to fusion, and from solar cells to how we smell.

Jo Locke is an accomplished educational consultant and author. She is known for her contributions to science education, particularly in developing engaging and effective learning resources for students and teachers worldwide.

Dr Douglas Palmer is an Earth science writer with over 20 published books. He has a background in academic palaeontology at Trinity College, University of Dublin.

Giles Sparrow is an author and journalist specializing in astronomy and space exploration. He has written dozens of books, and is a Fellow of the Royal Astronomical Society.

CONTRIBUTORS

Jack Challoner is the author of more than 50 books on science and technology. Before becoming a writer, he worked at London's Science Museum. He studied physics and trained as a science and maths teacher.

Marek Walisiewicz studied life sciences at Oxford University and conservation at UCL. He is the author/editor of multiple books on science, technology, and natural history.

CHEMI

S T R Y

Chemistry is a science born of utility. Our ancestors strove to transform matter in order to improve their lives, so the first chemists were – arguably – the ancient peoples who fired clay, extracted dyes, and smelted metals. The first systematic studies of chemistry – including classifications of substances and early chemical equations – were the work of Arab alchemists and scholars. Today's chemists study the properties of substances and the nature of their reactions, seeking not only to comprehend the world of matter but to transform it at the molecular level. Studies in physics have helped to explain the chemical properties of matter at the atomic and subatomic levels.

THE BUILDING BLOCKS OF MATTER

Matter is anything that has mass and that occupies space (has volume). It is ultimately made up of quarks and leptons (see pp.74–75), which are known as elementary particles because they cannot be broken down into smaller units. These particles combine to form protons, neutrons, and electrons, which are the components of atoms. Chemistry is the study of matter and its interactions at the atomic level, describing how atoms combine to form molecules and compounds. Atoms are the basis of the elements (see p.12): one atom is the smallest unit of an element that still carries that element's properties.

Structure of an atom

At around ten billionths of a metre across, atoms are too small to be seen. Scientists have developed "models" of their structure that explain many of their characteristics. The basic model is that of an atomic nucleus containing protons and neutrons orbited by much lighter electrons.

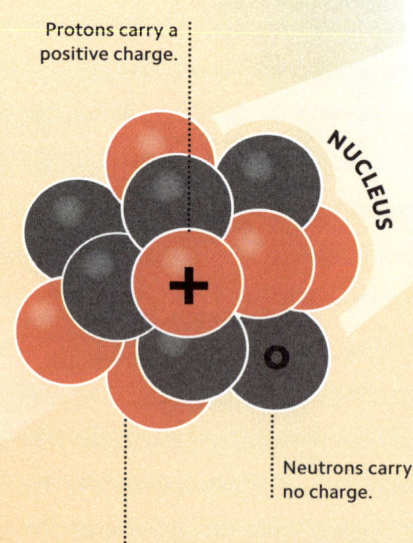

Protons carry a positive charge.

NUCLEUS

Neutrons carry no charge.

The atomic nucleus contains protons and neutrons. The number of protons determines the identity of the element. This atom is carbon, because it has six protons.

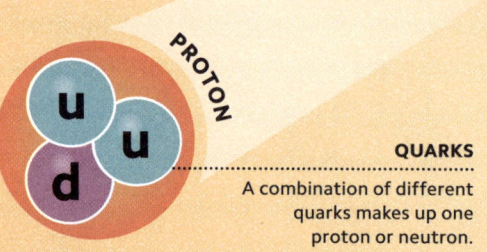

PROTON

QUARKS

A combination of different quarks makes up one proton or neutron.

NUCLEUS

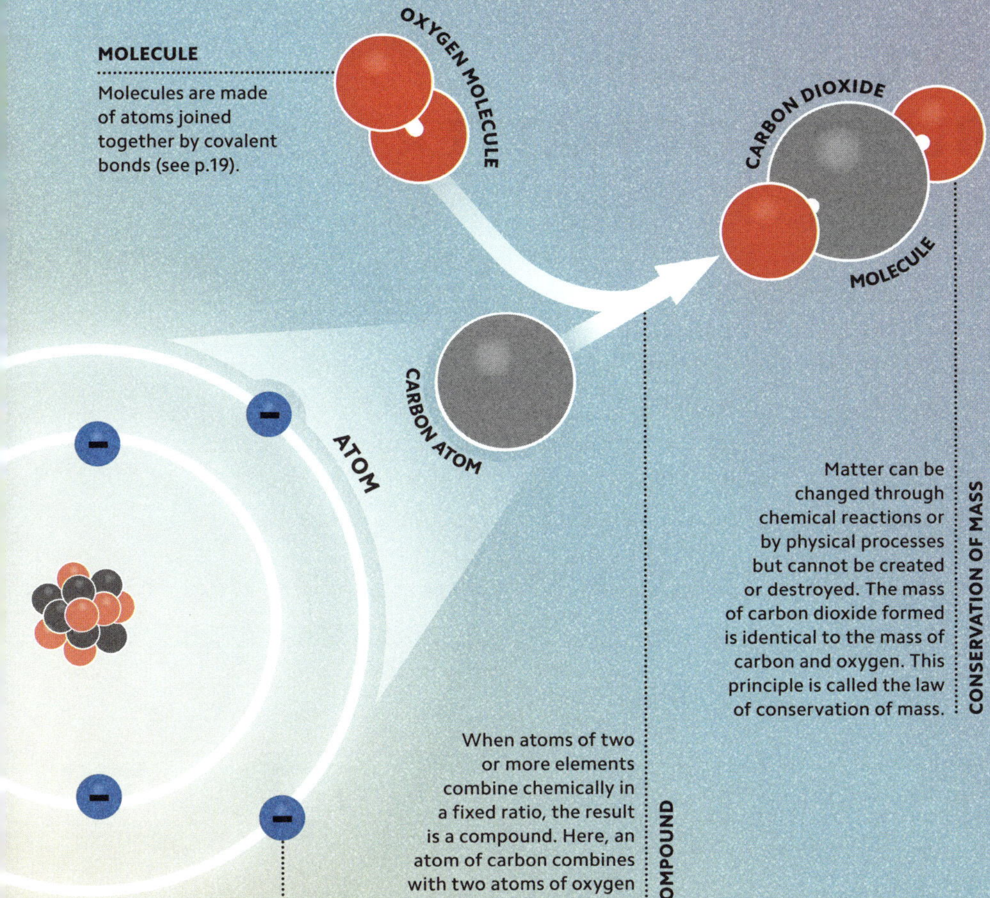

MOLECULE

Molecules are made of atoms joined together by covalent bonds (see p.19).

OXYGEN MOLECULE

CARBON ATOM

ATOM

CARBON DIOXIDE

MOLECULE

CONSERVATION OF MASS

Matter can be changed through chemical reactions or by physical processes but cannot be created or destroyed. The mass of carbon dioxide formed is identical to the mass of carbon and oxygen. This principle is called the law of conservation of mass.

COMPOUND

When atoms of two or more elements combine chemically in a fixed ratio, the result is a compound. Here, an atom of carbon combines with two atoms of oxygen to form carbon dioxide.

ELECTRON

An electron is a type of lepton (see p.74) that carries a negative charge.

"No new creation or destruction of matter is within the reach of chemical agency."

John Dalton

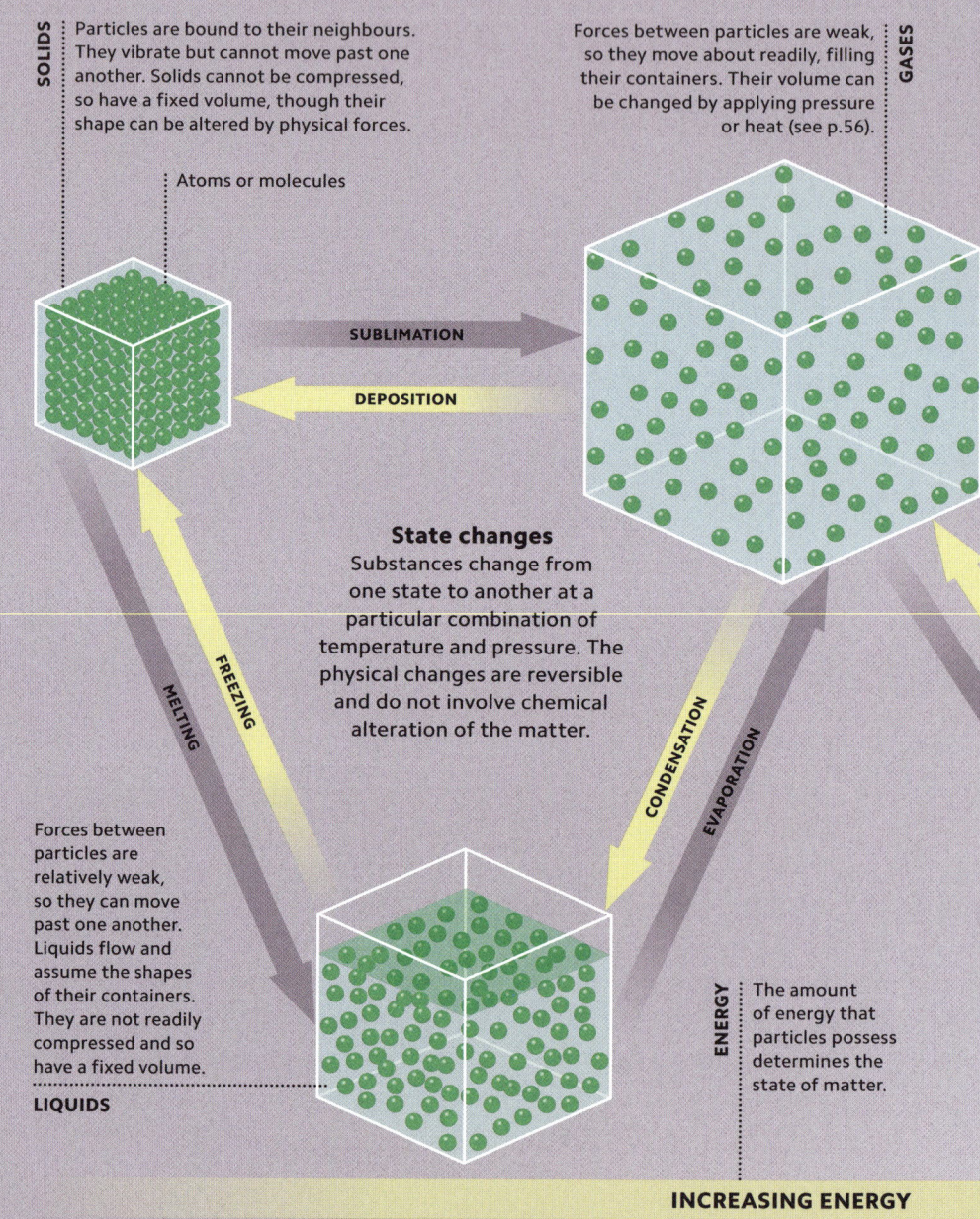

SOLIDS

Particles are bound to their neighbours. They vibrate but cannot move past one another. Solids cannot be compressed, so have a fixed volume, though their shape can be altered by physical forces.

Atoms or molecules

GASES

Forces between particles are weak, so they move about readily, filling their containers. Their volume can be changed by applying pressure or heat (see p.56).

SUBLIMATION

DEPOSITION

State changes

Substances change from one state to another at a particular combination of temperature and pressure. The physical changes are reversible and do not involve chemical alteration of the matter.

FREEZING

MELTING

CONDENSATION

EVAPORATION

Forces between particles are relatively weak, so they can move past one another. Liquids flow and assume the shapes of their containers. They are not readily compressed and so have a fixed volume.

LIQUIDS

ENERGY

The amount of energy that particles possess determines the state of matter.

INCREASING ENERGY

MOVING PARTICLES, CHANGING STATES

Even though matter appears to be static, its constituent particles are in constant random motion. The energy of movement (kinetic energy) that they possess, together with the strength of the bonds between particles, determines the state of the system – that is, whether it is a solid, liquid, gas, or plasma. Adding heat energy causes the particles to move faster – sometimes fast enough to overcome the bonds holding them in place. This is why heating a solid will eventually cause it to melt, and heating a liquid will cause it to evaporate. Adding energy to a gas may strip electrons off its component atoms or molecules, creating a plasma – a fourth state of matter – seen as flames, lightning, auroras, and in stars.

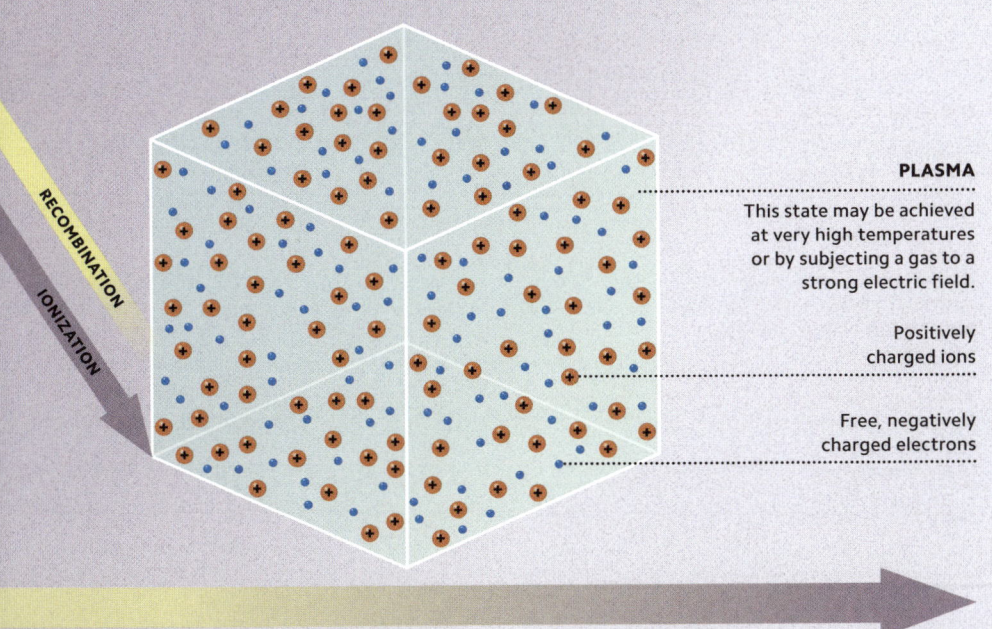

RECOMBINATION

IONIZATION

PLASMA
This state may be achieved at very high temperatures or by subjecting a gas to a strong electric field.

Positively charged ions

Free, negatively charged electrons

ATOMS BY NUMBERS

An element is a substance made up of just one type of atom. It cannot be made into a simpler substance by chemical means. The atoms of each element are distinct, having a characteristic number of protons in their nucleus. A hydrogen atom, for example, has just one proton in its nucleus; carbon has six, and copper 29. The number of protons in a neutral atom is matched by the number of electrons around its nucleus, and it is this number that determines the types of chemical bonds which that element makes. The nucleus of an element also contains neutrons, which contribute to the mass of the element but not its chemical nature.

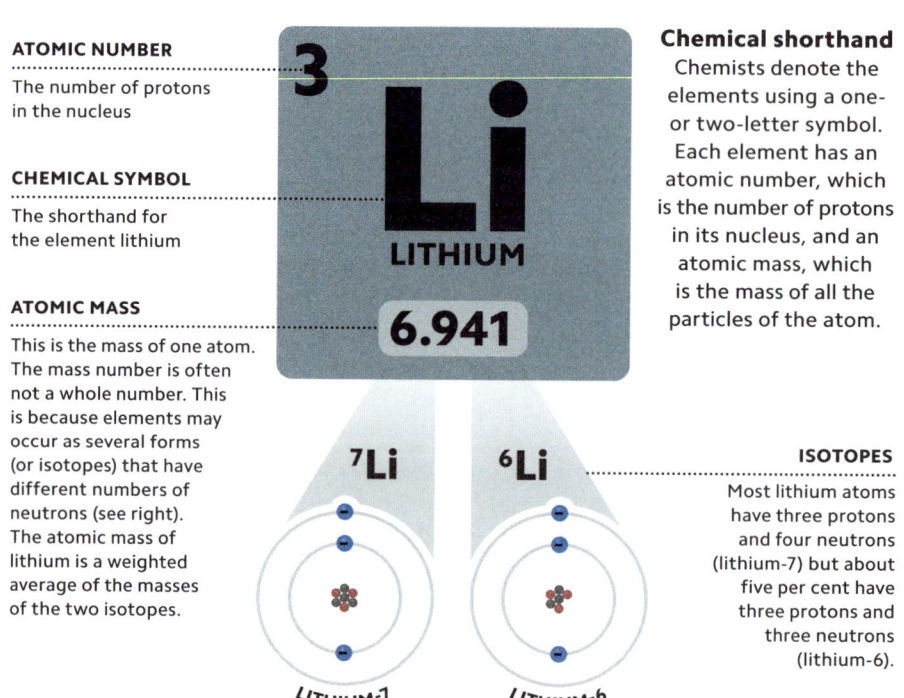

ATOMIC NUMBER

The number of protons in the nucleus

CHEMICAL SYMBOL

The shorthand for the element lithium

ATOMIC MASS

This is the mass of one atom. The mass number is often not a whole number. This is because elements may occur as several forms (or isotopes) that have different numbers of neutrons (see right). The atomic mass of lithium is a weighted average of the masses of the two isotopes.

3

Li

LITHIUM

6.941

Chemical shorthand

Chemists denote the elements using a one- or two-letter symbol. Each element has an atomic number, which is the number of protons in its nucleus, and an atomic mass, which is the mass of all the particles of the atom.

^{7}Li

^{6}Li

LITHIUM-7

LITHIUM-6

ISOTOPES

Most lithium atoms have three protons and four neutrons (lithium-7) but about five per cent have three protons and three neutrons (lithium-6).

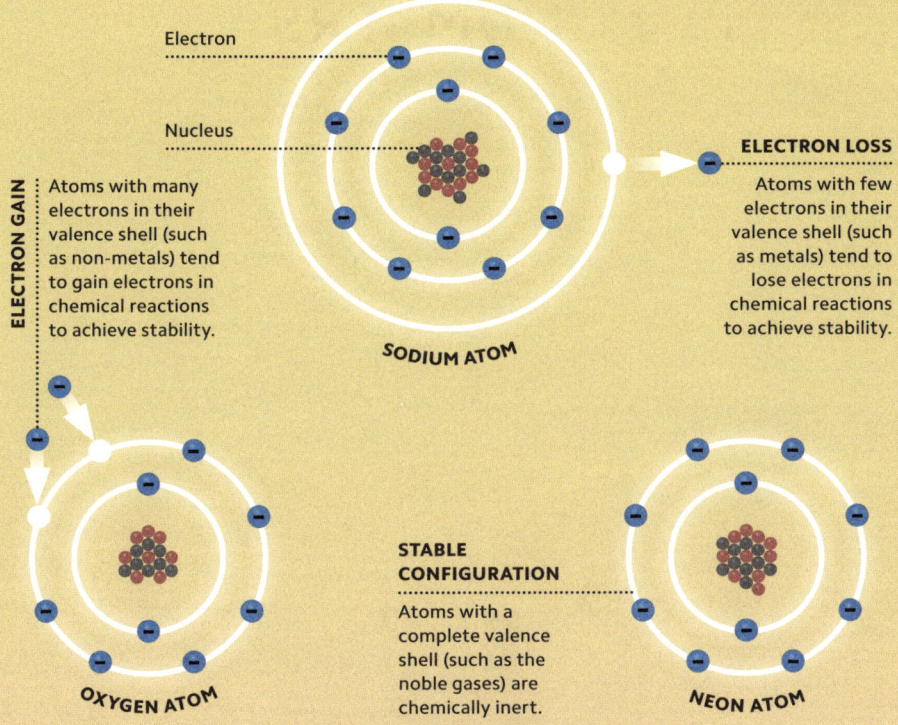

Electron

Nucleus

ELECTRON LOSS
Atoms with few electrons in their valence shell (such as metals) tend to lose electrons in chemical reactions to achieve stability.

ELECTRON GAIN
Atoms with many electrons in their valence shell (such as non-metals) tend to gain electrons in chemical reactions to achieve stability.

SODIUM ATOM

STABLE CONFIGURATION
Atoms with a complete valence shell (such as the noble gases) are chemically inert.

OXYGEN ATOM

NEON ATOM

ELECTRONS IN ORBIT

The electrons in an atom can be thought of as circling the nucleus in concentric orbits or "shells". The first shell of an atom can accommodate just two electrons; the next holds up to eight; the third 18. Each shell represents an energy level; electrons in shells close to the nucleus have least energy, those in the outermost (valence) shell have the most. Atoms are most stable (or least reactive) when their valence shell is full. To achieve a full valence shell, atoms can either lose or gain electrons during chemical reactions. The number of electrons in the valence shell determines the types of bonds that the atom makes, and therefore its chemical characteristics.

KEY

- **Hydrogen** – a colourless gas

REACTIVE METALS

- **Alkali metals** – soft, reactive metals
- **Alkaline earth metals** – moderately reactive metals

TRANSITION ELEMENTS

- **Transition metals** – a varied group, many of which have valuable properties

MAINLY NON-METALS

- **Metalloids** – possess properties between those of metals and non-metals
- **Other metals** – mostly relatively soft metals with low melting points
- **Carbon** and other non-metals
- **Halogens** – very reactive non-metals
- **Noble gases** – colourless, unreactive gases

RARE EARTH METALS

- Also called lanthanides and actinides, these are reactive metals. Some are rare or synthetic.

Atomic number

Atomic mass

GROUPS

Each column of the table is called a group. Elements are ordered by atomic number from top to bottom of a column. The atoms of all the elements in a group have the same number of electrons in their outermost shell (see p.13) and so have similar chemical properties.

PERIODS

Each row of the table is called a period. Elements are ordered by atomic number from left to right within a period. The number of electrons in the outermost shell of the atoms increases from left to right. All the elements in a period have the same number of electron shells.

1						
1 **H** HYDROGEN 1.008	2					
3 **Li** LITHIUM 6.941	**4** **Be** BERYLLIUM 9.0122					
11 **Na** SODIUM 22.990	**12** **Mg** MAGNESIUM 24.305	3	4	5	6	7
19 **K** POTASSIUM 39.098	**20** **Ca** CALCIUM 40.078	**21** **Sc** SCANDIUM 44.956	**22** **Ti** TITANIUM 47.867	**23** **V** VANADIUM 50.942	**24** **Cr** CHROMIUM 51.996	**25** **Mn** MANGANESE 54.938
37 **Rb** RUBIDIUM 85.468	**38** **Sr** STRONTIUM 87.62	**39** **Y** YTTRIUM 88.906	**40** **Zr** ZIRCONIUM 91.224	**41** **Nb** NIOBIUM 92.906	**42** **Mo** MOLYBDENUM 95.96	**43** **Tc** TECHNETIUM (98)
55 **Cs** CAESIUM 132.91	**56** **Ba** BARIUM 137.33	57–71	**72** **Hf** HAFNIUM 178.49	**73** **Ta** TANTALUM 180.95	**74** **W** TUNGSTEN 183.84	**75** **Re** RHENIUM 186.21
87 **Fr** FRANCIUM (223)	**88** **Ra** RADIUM (226)	89–103	**104** **Rf** RUTHERFORDIUM (267)	**105** **Db** DUBNIUM (268)	**106** **Sg** SEABORGIUM (269)	**107** **Bh** BOHRIUM (270)

These metallic elements, some radioactive and synthetic, are by convention shown at the bottom of the periodic table because they would greatly increase the width of the table if included in their correct locations.

LANTHANIDES AND ACTINIDES

57 **La** LANTHANUM 138.91	**58** **Ce** CERIUM 140.12	**59** **Pr** PRAESODYMIUM 140.91	**60** **Nd** NEODYMIUM 144.24
89 **Ac** ACTINIUM (227)	**90** **Th** THORIUM 232.04	**91** **Pa** PROTACTINIUM 231.04	**92** **U** URANIUM 238.03

CATALOGUING THE ELEMENTS

Chemists in the 19th century saw that different elements had similar characteristics – for example, they could be metallic or non -metallic, reactive or less reactive, and make similar types of compounds. They used these similarities to place elements in groups. When ordering the elements by their atomic number, they further realized that these similarities recurred at definite intervals – rather like musical notes in successive octaves, they showed periodicity. In the 1860s the Russian chemist Dmitri Mendeléev proposed his periodic table of the elements.

8	9	10	11	12	13	14	15	16	17	18
										2 He HELIUM 4.003
					5 B BORON 10.81	6 C CARBON 12.011	7 N NITROGEN 14.007	8 O OXYGEN 15.999	9 F FLUORINE 18.998	10 Ne NEON 20.180
					13 Al ALUMINIUM 26.982	14 Si SILICON 28.085	15 P PHOSPHORUS 30.974	16 S SULFUR 32.06	17 Cl CHLORINE 35.45	18 Ar ARGON 39.948
26 Fe IRON 55.845	27 Co COBALT 58.933	28 Ni NICKEL 58.693	29 Cu COPPER 63.546	30 Zn ZINC 65.38	31 Ga GALLIUM 69.723	32 Ge GERMANIUM 72.63	33 As ARSENIC 74.922	34 Se SELENIUM 78.971	35 Br BROMINE 79.904	36 Kr KRYPTON 83.798
44 Ru RUTHENIUM 101.07	45 Rh RHODIUM 102.91	46 Pd PALLADIUM 106.42	47 Ag SILVER 107.87	48 Cd CADMIUM 112.41	49 In INDIUM 114.82	50 Sn TIN 118.71	51 Sb ANTIMONY 121.76	52 Te TELLURIUM 127.60	53 I IODINE 126.90	54 Xe XENON 131.29
76 Os OSMIUM 190.23	77 Ir IRIDIUM 192.22	78 Pt PLATINUM 195.08	79 Au GOLD 196.97	80 Hg MERCURY 200.59	81 Tl THALLIUM 204.38	82 Pb LEAD 207.20	83 Bi BISMUTH 208.98	84 Po POLONIUM (209)	85 At ASTATINE (210)	86 Rn RADON (222)
108 Hs HASSIUM (277)	109 Mt MEITNERIUM (278)	110 Ds DARMSTADTIUM (281)	111 Rg ROENTGENIUM (282)	112 Cn COPERNICIUM (285)	113 Nh NIHONIUM (286)	114 Fl FLEROVIUM (289)	115 Mc MOSCOVIUM (289)	116 Lv LIVERMORIUM (293)	117 Ts TENNESSINE (294)	118 Og OGANESSON (294)

61 Pm PROMETHIUM (145)	62 Sm SAMARIUM 150.36	63 Eu EUROPIUM 151.96	64 Gd GADOLINIUM 157.25	65 Tb TERBIUM 158.93	66 Dy DYSPROSIUM 162.50	67 Ho HOLMIUM 164.93	68 Er ERBIUM 167.26	69 Tm THULIUM 168.93	70 Yb YTTERBIUM 173.05	71 Lu LUTETIUM 174.97
93 Np NEPTUNIUM (237)	94 Pu PLUTONIUM (244)	95 Am AMERICIUM (243)	96 Cm CURIUM (247)	97 Bk BERKELIUM (247)	98 Cf CALIFORNIUM (251)	99 Es EINSTEINIUM (252)	100 Fm FERMIUM (257)	101 Md MENDELEVIUM (258)	102 No NOBELIUM (259)	103 Lr LAWRENCIUM (262)

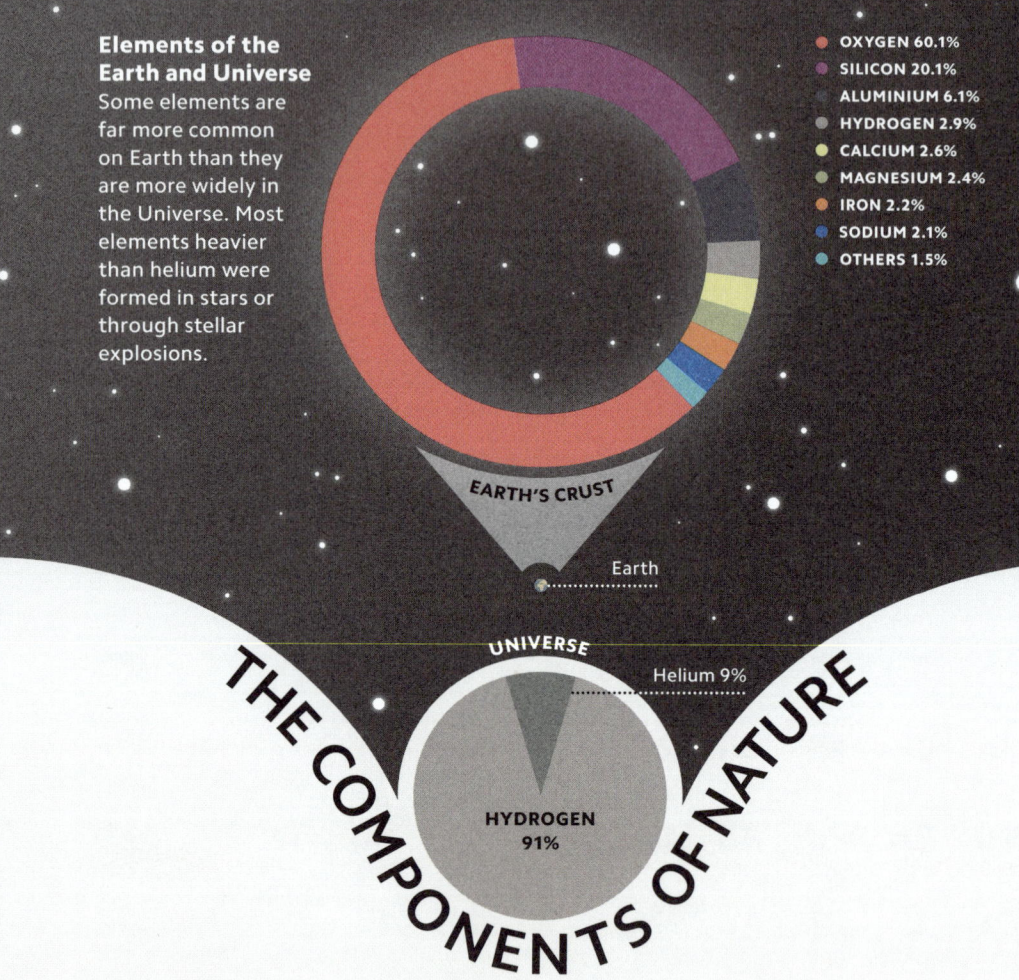

Elements of the Earth and Universe
Some elements are far more common on Earth than they are more widely in the Universe. Most elements heavier than helium were formed in stars or through stellar explosions.

- ● OXYGEN 60.1%
- ● SILICON 20.1%
- ● ALUMINIUM 6.1%
- ● HYDROGEN 2.9%
- ● CALCIUM 2.6%
- ● MAGNESIUM 2.4%
- ● IRON 2.2%
- ● SODIUM 2.1%
- ● OTHERS 1.5%

EARTH'S CRUST

Earth

UNIVERSE

Helium 9%

HYDROGEN 91%

THE COMPONENTS OF NATURE

Scientists today recognize 118 elements, 94 of which occur naturally and 24 of which are synthetic, existing for very short periods as the products of nuclear manipulation. A few elements, such as gold and silver, occur in nature in their pure (native) form, but the majority are chemically combined with other substances. Most elements have cosmic origins; some formed at the birth of the Universe, others during the lives and deaths of stars.

ELEMENTAL DIVIDE

Some three-quarters of known elements are metals, which are found at the left and centre of the periodic table (see pp.14–15). Their atoms have 1–3 electrons in their outermost shell and tend to form compounds by losing these outer electrons (see p.18). Non-metals, which sit to the right of the periodic table, have 4–7 electrons in their outermost shell and tend to make compounds by gaining electrons in their outer shell (see p.19). Non-metals include the so-called noble gases – such as helium, neon, and argon – which are highly unreactive because they have a full complement of electrons in their outer shell.

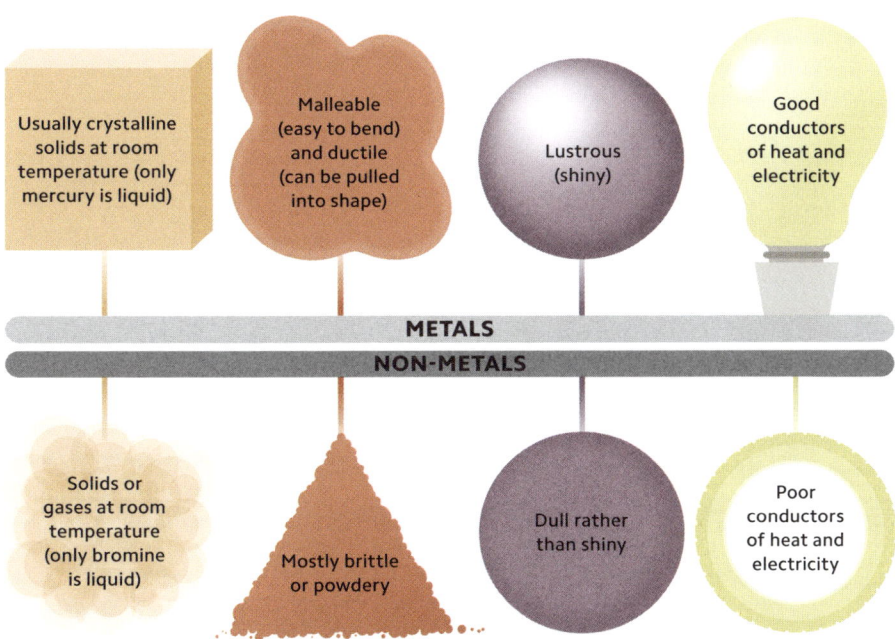

Usually crystalline solids at room temperature (only mercury is liquid)

Malleable (easy to bend) and ductile (can be pulled into shape)

Lustrous (shiny)

Good conductors of heat and electricity

METALS

NON-METALS

Solids or gases at room temperature (only bromine is liquid)

Mostly brittle or powdery

Dull rather than shiny

Poor conductors of heat and electricity

Sodium loses an electron

Chlorine gains an electron

SODIUM ATOM: Na

CHLORINE ATOM: Cl

CHARGED ATTRACTION

Atoms can make bonds with other atoms. The nature of the bond depends on the atoms involved, and it determines the properties of the resulting compound. Metal atoms have few electrons in their valence shell (see p.13), while non-metal atoms have shells that are almost complete. In one type of bond, the valence electrons of a metal atom are transferred to the valence shell of a non-metal. This is known as an ionic bond and it depends on electrostatic attraction.

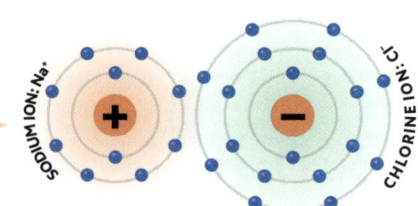

SODIUM ION: Na⁺

CHLORINE ION: Cl⁻

SODIUM CHLORIDE: NaCl

Forming an ionic bond

A sodium atom will readily give up its single valence electron and chlorine will readily accept the electron. The result is two oppositely charged ions held together by strong forces. Most ionic compounds are crystalline solids at room temperature and have high melting and boiling points.

Forming a covalent bond

Ammonia is a covalent compound formed when three hydrogen atoms share their single electron with the three valence electrons of a nitrogen atom. The bonds within covalent molecules are strong, but the attraction between molecules is weak, so these compounds have low melting and boiling points. Some are gases at room temperature.

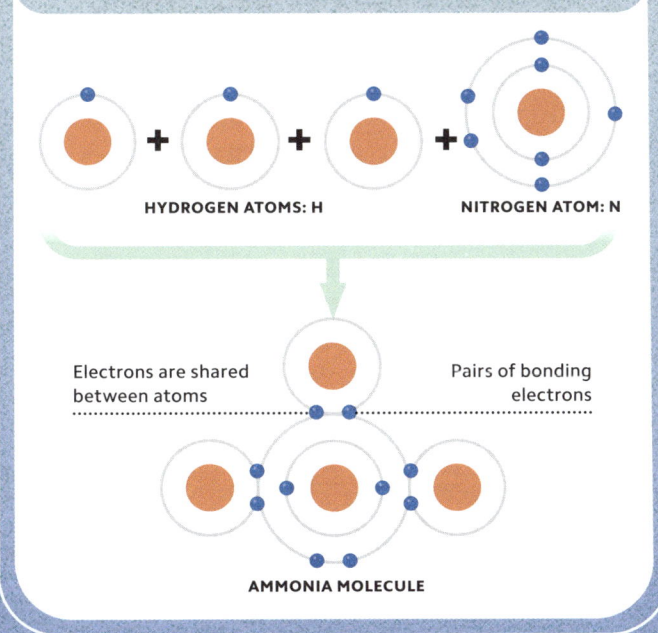

HYDROGEN ATOMS: H

NITROGEN ATOM: N

Electrons are shared between atoms

Pairs of bonding electrons

AMMONIA MOLECULE

SHARED ATTRACTION

While ionic compounds gain their strength by transferring electrons between atoms, a different type of bonding is seen in non-metallic elements. This is called covalent bonding, and it relies on the sharing, rather than transfer, of electrons. Covalent bonds can form between atoms of one element or between atoms of different elements.

SEAS OF ELECTRONS

The atoms that make up metals are held together by metallic bonds. Like ionic bonds (see p.18), these rely on electrostatic forces to keep the atoms packed closely together in a regular, three-dimensional array, or crystal. Iron and aluminium, for example, form cubic crystals. Metal atoms have few electrons in their valence shell (see p.13), and these electrons are only loosely bound to the nucleus. They tend to detach from their parent atom, forming a "sea" of negative charge that glues together positively charged metal ions.

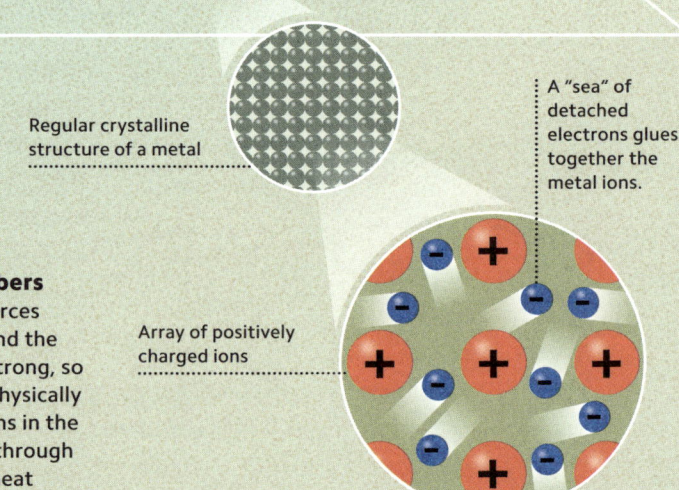

Regular crystalline structure of a metal

A "sea" of detached electrons glues together the metal ions.

Strength in numbers

The electrostatic forces between the ions and the free electrons are strong, so metals are in turn physically strong. The electrons in the "sea" move readily through the solid, carrying heat energy or an electric current.

Array of positively charged ions

Carbon has four electrons in its valence shell, so can make four covalent bonds. This means that it can link together with other carbon atoms to form long chain polymers and biomolecules (see pp.100–101). Here, carbon atoms form the backbone of glycine, an amino acid.

FOUR BONDS

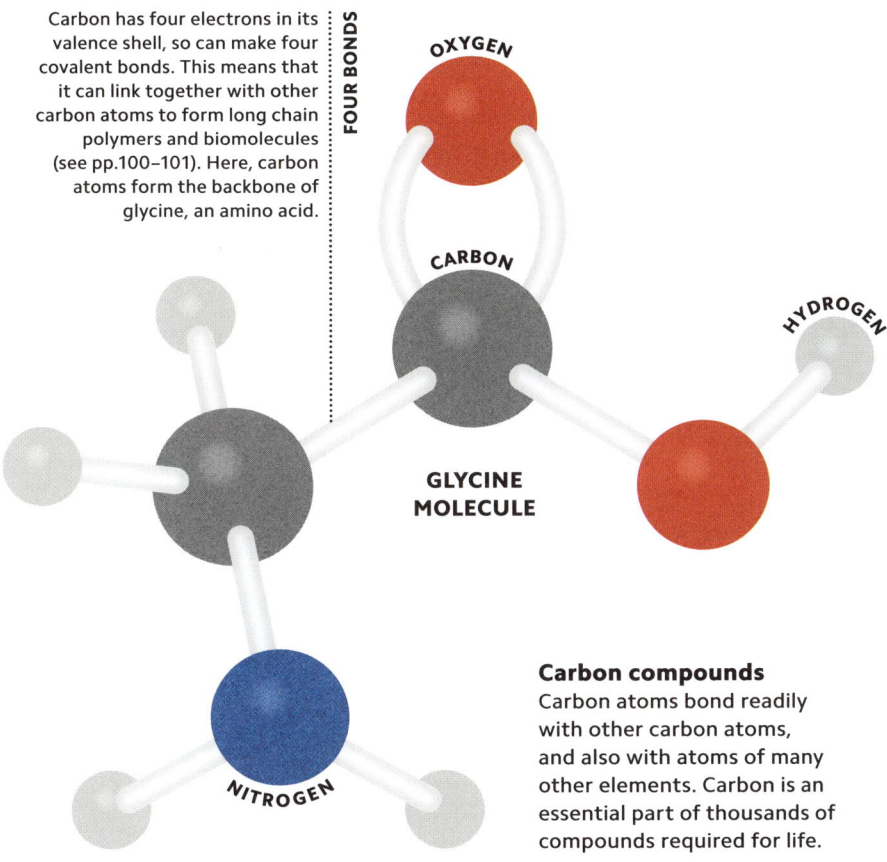

OXYGEN

CARBON

HYDROGEN

GLYCINE MOLECULE

NITROGEN

Carbon compounds

Carbon atoms bond readily with other carbon atoms, and also with atoms of many other elements. Carbon is an essential part of thousands of compounds required for life.

THE CHEMISTRY OF LIFE

All living things on our planet depend on the chemistry of a handful of elements, notably carbon and hydrogen, but also oxygen, nitrogen, sulfur, and phosphorus. The central importance of carbon to life has established the study of its compounds as a branch of chemistry called organic chemistry (though not all carbon compounds are natural).

MIXED UP

The constituent atoms of a compound are bound to one another by chemical bonds, which may be ionic, covalent, or metallic (se pp.18–20). In contrast, the components of a mixture are not chemically bonded to one another and can be separated by purely physical processes. Mixtures can be between solids, liquids, or gases. They possess the properties of all the constituents, while a compound has a set of physical and chemical properties all of its own.

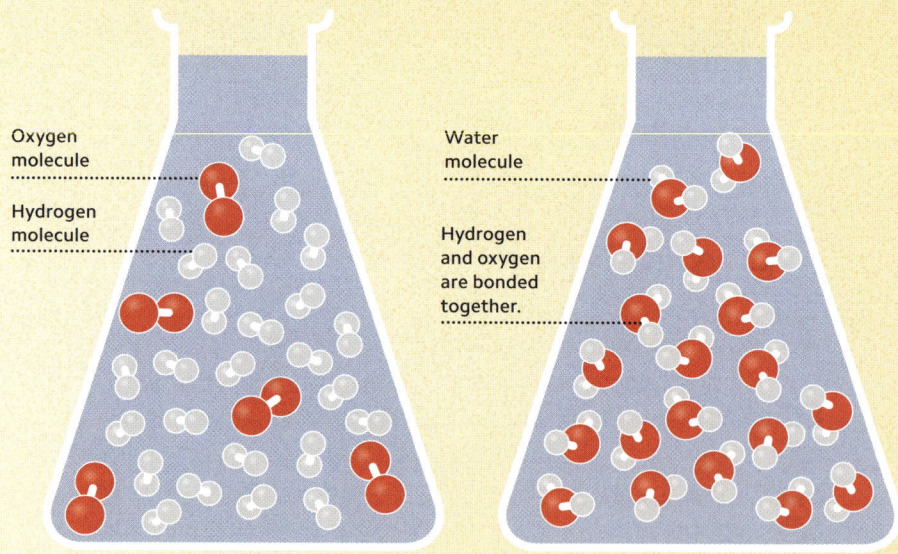

Oxygen molecule

Hydrogen molecule

Water molecule

Hydrogen and oxygen are bonded together.

Mixture
A mixture can have any ratio of constituents. Here, for example, there are five molecules of hydrogen to each molecule of oxygen, but there could be many more or many fewer.

Compound
The constituent atoms of a compound are always present in exactly the same proportions; so, for example, any sample of water (H_2O) will contain twice the number of hydrogen atoms as oxygen atoms.

SLIGHT NEGATIVE CHARGE

The oxygen atom in a water molecule is said to be electronegative – it exerts a strong pull on the electrons in the molecule. The oxygen atom thus carries a slight negative charge.

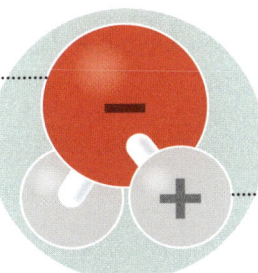

WATER MOLECULE

SLIGHT POSITIVE CHARGE

The hydrogen atoms are left with a slight positive charge.

The "universal" solvent

Water is an excellent solvent that can dissolve many compounds, including ionic compounds such as common salt (sodium chloride). It can do this because water molecules have a slight asymmetry in their electric charge.

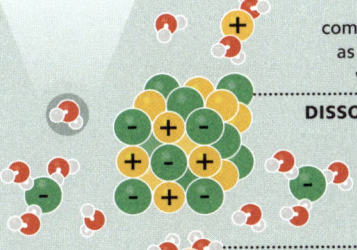

Most ionic compounds, such as salt, dissolve well in water.

DISSOLVING SALT CRYSTAL

WATER MOLECULE

The slight negative charge of oxygen atoms stabilizes the positively charged ions and the slight positive charge of hydrogen atoms stabilizes negatively charged ions.

IN SEARCH OF A SOLUTION

A solution is a close mixture of two or more substances in which the particles of one substance (the solvent) surround the particles of the other (the solute). Gases and solids both form solutions – air is a gaseous solution (see p.25) and metal alloys are solid solutions – but the most familiar solutions are liquids. Salt water is a solution that makes up the seas and oceans. Solutions form when the attraction between solute and solvent particles is similar to, or greater than, the attractions between solute particles.

WATER WORLD

Most substances get denser as they freeze (change state from liquid to solid). Water, however, becomes less dense. This means that ice forms on the surface of bodies of water, allowing living things to remain unfrozen below.

DENSITY

Water has a high specific heat capacity, meaning that a lot of energy is needed to make its temperature rise. This makes water a stable environment in which temperature fluctuations are generally low.

HEAT CAPACITY

Water is the most abundant molecule on our planet. It makes up over 60 per cent of our bodies and is one of the few compounds to exist as a solid, liquid, and gas on Earth's surface. The molecule has a number of properties, including its ability to dissolve a wide range of substances (see p.23), that help to sustain life on our planet.

Water molecules have "poles" because they carry an imbalance of charge (see p.23). This makes them good solvents, and also means that water molecules are "sticky" – attracted to one another. For this reason, water has relatively high boiling and melting points.

POLARITY

The tendency of water molecules to "stick" together and to other molecules is vital to life. Trees offer a good example. The water needed by the leaves of a tall tree rises by capillary action – by "sticking" to the walls of long, thin tubules in the tree trunk.

ADHESION

OXYGEN
20.95%

OTHER GASES
These include
carbon dioxide
(0.038 per cent)
and traces of
others such
as helium,
neon, hydrogen,
and krypton.

NITROGEN
78.08%

Composition of air
Air is a solution in which
the main component (the
solvent) is nitrogen and
the solutes are oxygen
and other gases present
in small amounts.

A SEA OF GASES

The mixture of
gases that makes up our
atmosphere is colourless and
odourless. Its composition remains the
same from sea level to altitudes of around
100 km (60 miles), though its density decreases.
The gases in air have not been constant over time.
Billions of years ago, volcanic activity produced an
unbreathable atmosphere rich in carbon monoxide and
dioxide, sulfur compounds, water vapour, chlorine, and
ammonia. The evolution of photosynthesis (see p.103)
and the breakdown of gases by sunlight helped
create the air we breathe today.

GETTING A REACTION

Types of reaction
Chemical reactions are expressed as equations. These must balance – that is, have the same number of atoms on either side of the arrow (see image below) – because matter cannot be created or destroyed in a chemical reaction. There are several basic types of reaction.

Elements or compounds can be converted into other chemical forms through reactions. These are processes that involve breaking or making chemical bonds, whether covalent, ionic, or metallic (see pp.18–20). The substances that react together are called reactants, and the new substances that form are called products. Some reactions occur spontaneously, while others need inputs of energy, pressure, or the presence of another substance called a catalyst (see p.30).

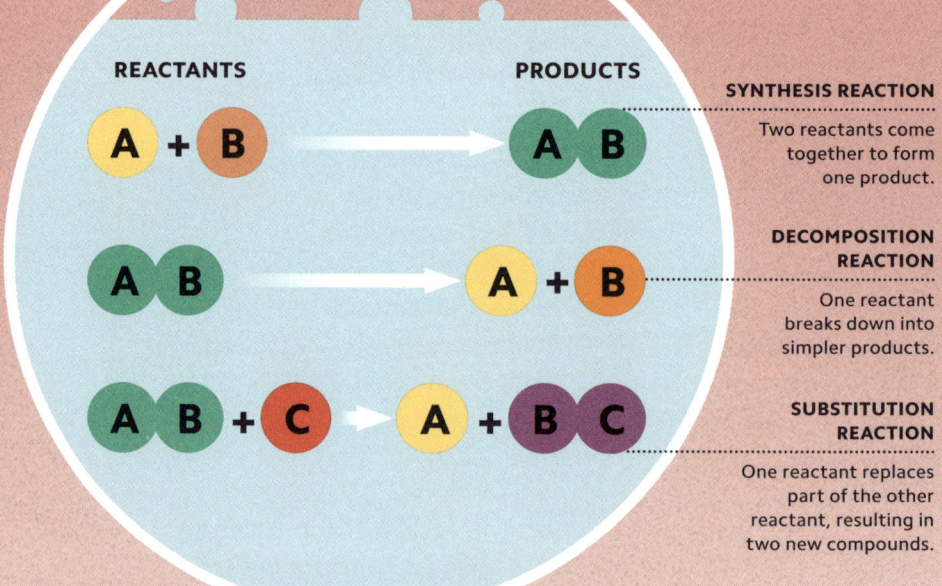

REACTANTS   PRODUCTS

A + B → A B

SYNTHESIS REACTION
Two reactants come together to form one product.

A B → A + B

DECOMPOSITION REACTION
One reactant breaks down into simpler products.

A B + C → A + B C

SUBSTITUTION REACTION
One reactant replaces part of the other reactant, resulting in two new compounds.

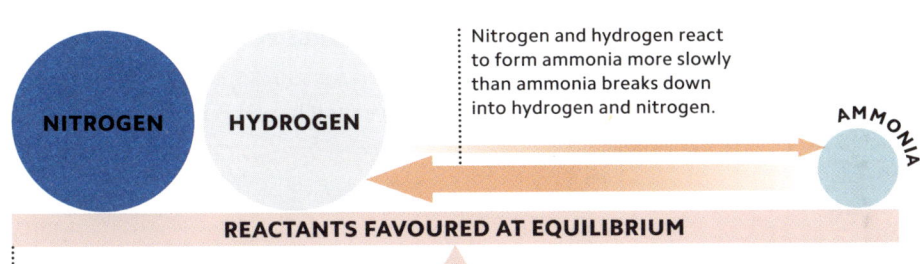

Nitrogen and hydrogen react to form ammonia more slowly than ammonia breaks down into hydrogen and nitrogen.

REACTANTS FAVOURED AT EQUILIBRIUM

An equilibrium is established in which there is much more of the reactants than the products in the reaction mix.

IN THE BALANCE

The reactions that occur when a match burns cannot be reversed; it is impossible to make a match from ash. However, many reactions are reversible and – rather than being unidirectional – proceed both forwards and backwards at the same time. In such reactions, a balance is eventually reached in which the ratio of reactant to product is stable. This is called equilibrium. The important chemical ammonia is made in a reversible reaction between hydrogen and nitrogen.

$$N_2 + 3H_2 \rightleftharpoons 2NH_3$$

NITROGEN   HYDROGEN   AMMONIA

The two arrows indicate a reversible reaction.

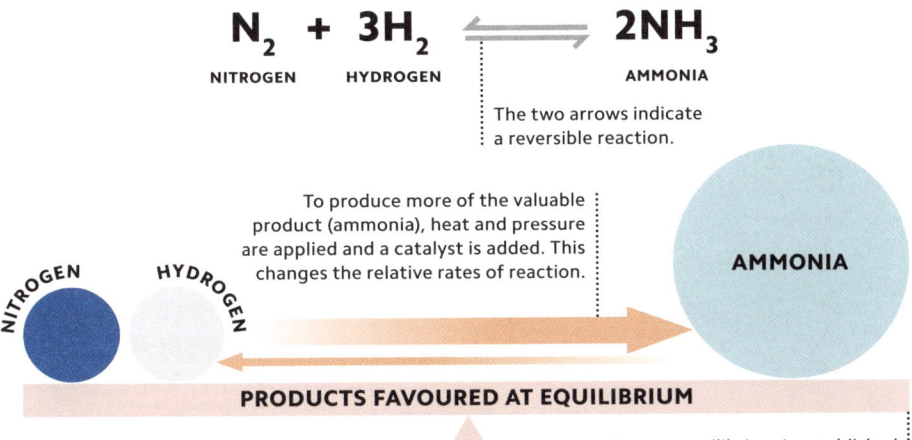

To produce more of the valuable product (ammonia), heat and pressure are applied and a catalyst is added. This changes the relative rates of reaction.

PRODUCTS FAVOURED AT EQUILIBRIUM

An new equilibrium is established in which there is much more of the product than of the reactants.

Activation energy

Whether a reaction is exo- or endothermic, energy needs to be applied to initiate the reaction. This energy "barrier" is called the energy of activation. Reactions that need little activation energy happen more readily than those with a high activation energy.

Heat in or out?

Reactions such as combustion and explosions are exothermic, releasing heat; endothermic reactions, such as cooking food, absorb heat from their environment.

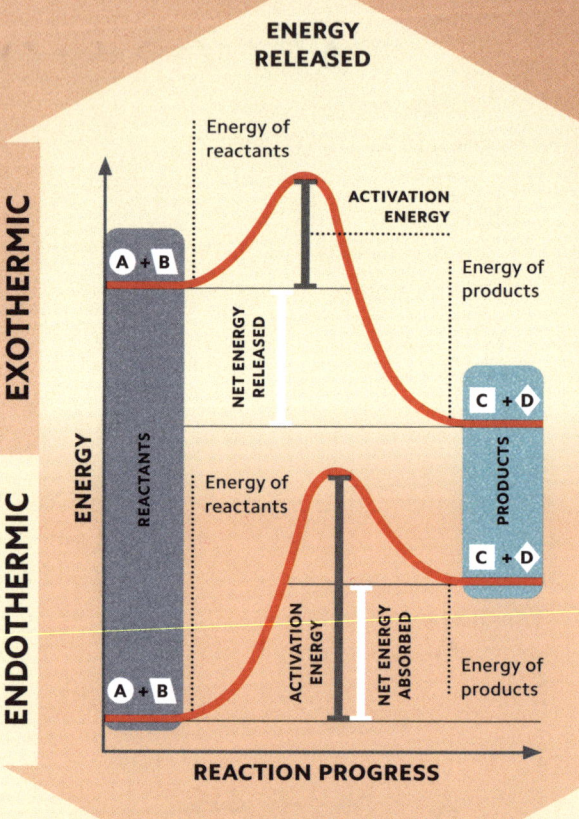

ENERGY RELEASED

Energy of reactants

ACTIVATION ENERGY

Energy of products

EXOTHERMIC

NET ENERGY RELEASED

ENERGY

REACTANTS

A + B

C + D

PRODUCTS

Energy of reactants

C + D

ENDOTHERMIC

ACTIVATION ENERGY

NET ENERGY ABSORBED

Energy of products

A + B

REACTION PROGRESS

ENERGY ABSORBED

ENERGETIC CHEMISTRY

Atoms "seek" greater stability by completing their valence shell of electrons (see p.13) and can achieve this by forming bonds with other atoms. When new bonds form, excess energy is released, usually in the form of heat. Conversely, when bonds are broken, energy is absorbed. Chemical reactions involve the making and/or breaking of bonds. Reactions that result in an overall release of heat energy are called exothermic, while those that absorb heat energy are endothermic.

FAST OR SLOW?

For a reaction to occur between two suitable molecules, they must first collide. What's more, the collision must occur with sufficient energy and with the molecules in the right orientation. Any factor that increases the speed (energy) of the molecules and the frequency of their collisions will make the reaction proceed faster. These factors include heat, which makes the molecules move faster; pressure (in gases) and concentration (in solutions), which bring molecules closer together; and catalysis, which holds molecules in the right orientation (see p.30).

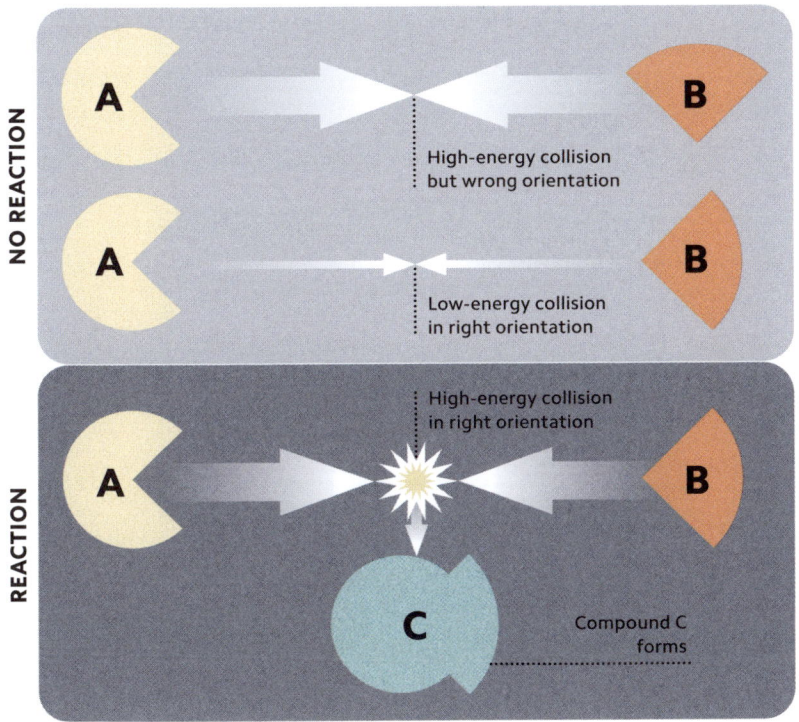

TAKING A NEW PATH

Many chemical reactions can be accelerated by adding a catalyst to the reactants. A catalyst is any substance that lowers the activation energy of a reaction (see p.28), but is not itself altered by the reaction. Metal catalysts, including iron, are widely used in industry; platinum and rhodium are used in catalytic converters in vehicles to reduce harmful emissions; and organic catalysts called enzymes (see p.99) are vital to chemical processes in living things.

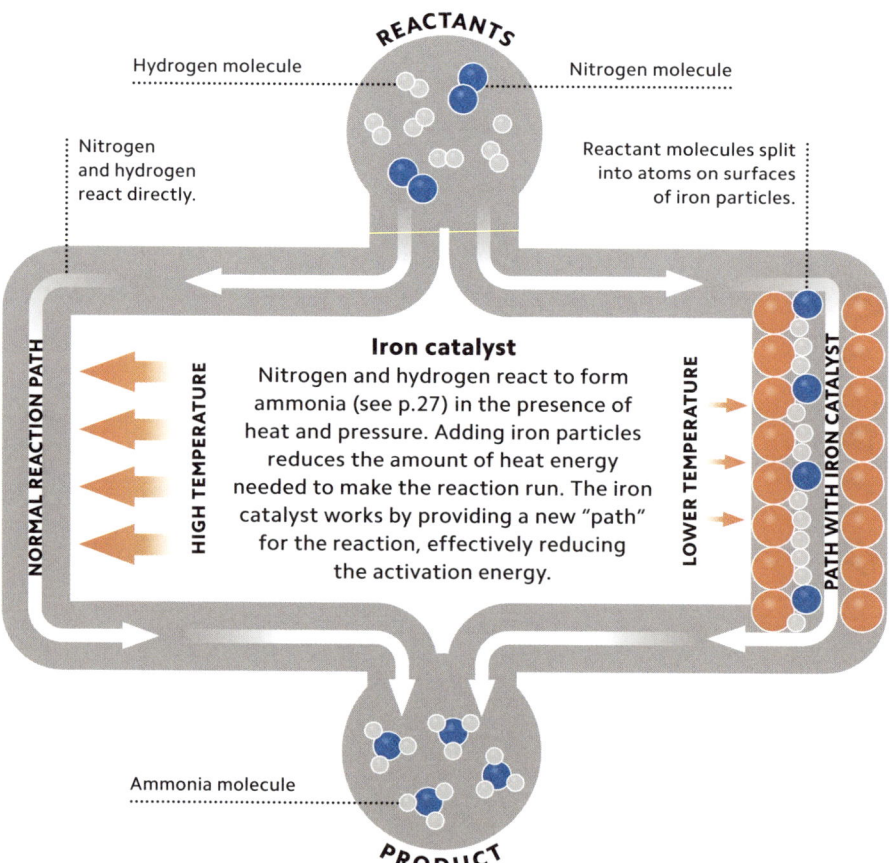

REACTANTS

Hydrogen molecule

Nitrogen molecule

Nitrogen and hydrogen react directly.

Reactant molecules split into atoms on surfaces of iron particles.

NORMAL REACTION PATH

HIGH TEMPERATURE

LOWER TEMPERATURE

PATH WITH IRON CATALYST

Iron catalyst
Nitrogen and hydrogen react to form ammonia (see p.27) in the presence of heat and pressure. Adding iron particles reduces the amount of heat energy needed to make the reaction run. The iron catalyst works by providing a new "path" for the reaction, effectively reducing the activation energy.

Ammonia molecule

PRODUCT

TNT
The explosive TNT actually contains less energy in its bonds than does petrol. It is so destructive because of the speed at which it releases its heat energy and the volume of gas produced as it decomposes.

DETONATION

TNT needs to be triggered to react (detonated) by applying heat energy or a pressure wave.

DECOMPOSITION

The bonds of the TNT break apart, releasing heat energy; this energy propagates at a very high speed, breaking more bonds and releasing more heat, and so on.

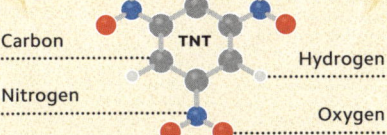

Carbon   TNT   Hydrogen

Nitrogen

Oxygen

BURNING AND BLOWING UP

Combustion is an exothermic (see p.28) reaction, in which a fuel reacts vigorously with oxygen (oxidizes) in the air, producing rapidly expanding gases. A spark or flame provides the activation energy (see p.28) to initiate the reaction. Small explosions may result if the combustion occurs in a sealed container, but the available oxygen is soon used up, stalling the reaction. Powerful explosives contain their own oxidizing chemicals, or – like TNT – rely on reactions other than combustion to produce energy.

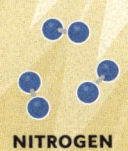

CARBON MONOXIDE

NITROGEN

Products
The gaseous products of the explosion expand extremely fast, giving TNT its explosive power.

WATER

CARBON

BATTERY ACID

APPLE JUICE

ACID RAIN

MILK

0 1 2 3 4 5 6 7

← INCREASINGLY ACIDIC

Measuring acidity

The acidity or alkalinity of a solution is measured on the pH scale, which ranges from 0–14. pH stands for the "potential (or power) of hydrogen".

FROM ACIDS TO BASES

Many compounds, both natural and synthetic, are classified as acids, bases, or salts. When bases are dissolved in water, they are known as alkalis. These classes of chemicals are used widely in industry, are found in essential domestic products, and are key to the chemistry of living things. The strength of acidity or alkalinity of a solution is measured on the pH scale – a measure of the concentration of hydrogen and hydroxide ions.

HYDROCHLORIC ACID (HCI)

Acids

When dissolved in water, acids release protons (hydrogen ions, H^+) into the solution. They are reactive (quickly corroding metals, for example) because the positively charged hydrogen ions exert a powerful "pull" on the negatively charged electrons of atoms that they encounter and so quickly make bonds.

Hydrogen ion (proton) (H^+)

Chlorine ion (Cl^-)

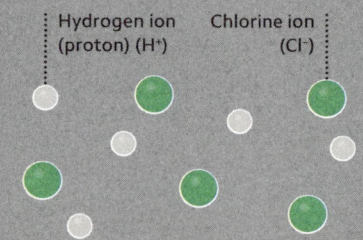

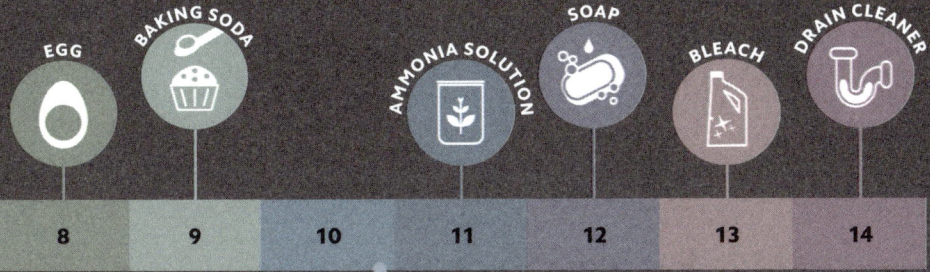

| 8 | 9 | 10 | 11 | 12 | 13 | 14 |

INCREASINGLY ALKALINE

SODIUM CHLORIDE + WATER
($NaCl + H_2O$)

Neutralization and salts

A salt is the product of a reaction between an acid and a base. When a base, such as sodium hydroxide, is added to a solution of an acid, such as hydrochloric acid, the protons (H^+) released by the acid react with the hydroxide ions (OH^-) released by the alkali to form neutral water (H_2O). The ions remaining in the solution are of dissolved sodium chloride.

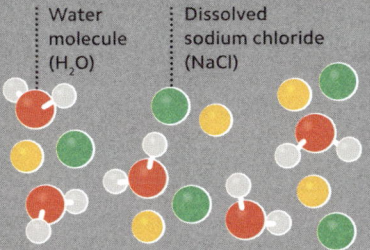

Water molecule (H_2O)

Dissolved sodium chloride (NaCl)

SODIUM HYDROXIDE
(NaOH)

Bases

These substances release hydroxide ions (OH^-) when dissolved in water. They are reactive because the hydroxide ions readily donate electrons to other atoms and so rapidly make bonds with them.

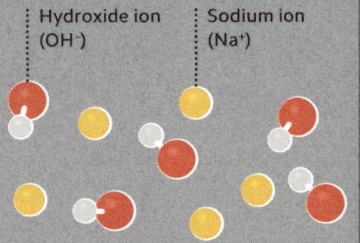

Hydroxide ion (OH^-)

Sodium ion (Na^+)

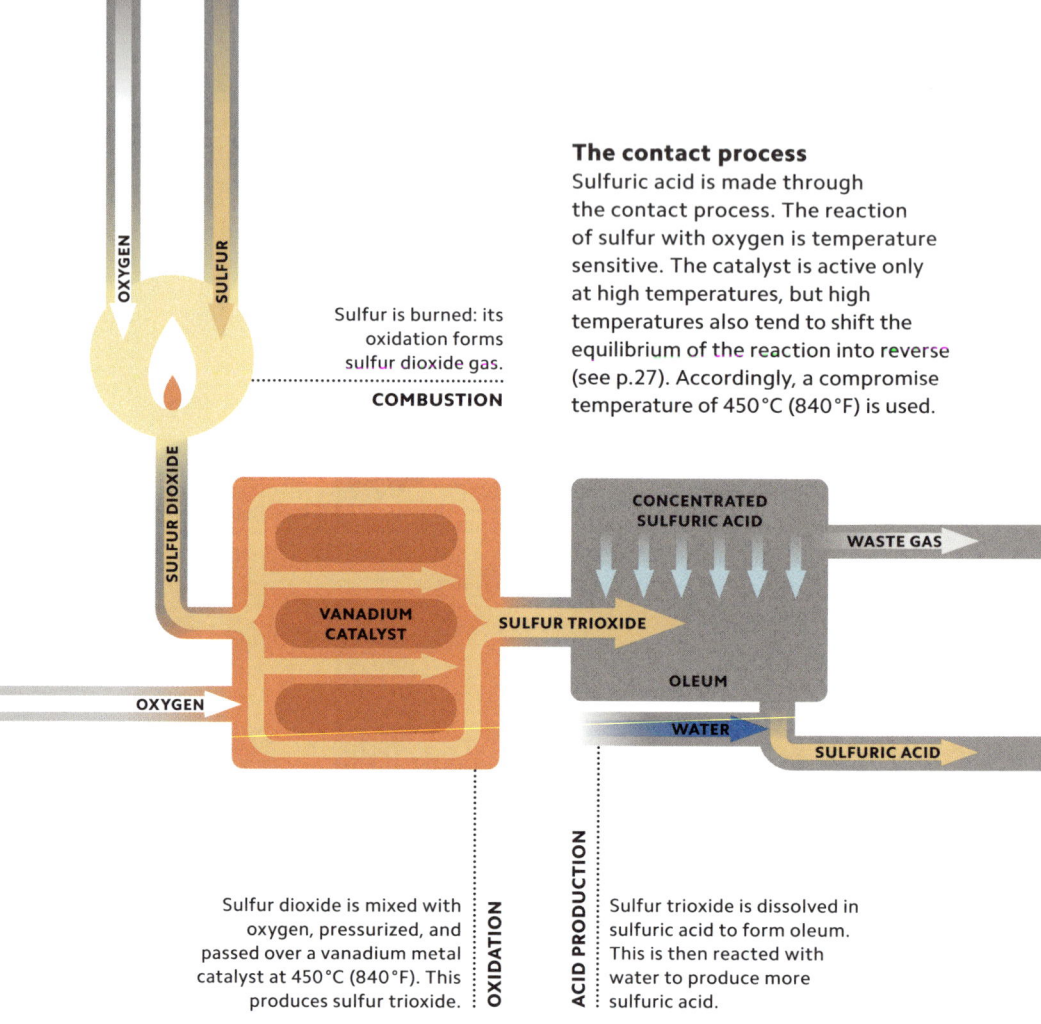

OXYGEN

SULFUR

Sulfur is burned: its
oxidation forms
sulfur dioxide gas.

COMBUSTION

The contact process
Sulfuric acid is made through
the contact process. The reaction
of sulfur with oxygen is temperature
sensitive. The catalyst is active only
at high temperatures, but high
temperatures also tend to shift the
equilibrium of the reaction into reverse
(see p.27). Accordingly, a compromise
temperature of 450°C (840°F) is used.

SULFUR DIOXIDE

CONCENTRATED
SULFURIC ACID

WASTE GAS

**VANADIUM
CATALYST**

SULFUR TRIOXIDE

OLEUM

OXYGEN

WATER

SULFURIC ACID

Sulfur dioxide is mixed with
oxygen, pressurized, and
passed over a vanadium metal
catalyst at 450°C (840°F). This
produces sulfur trioxide.

OXIDATION

ACID PRODUCTION

Sulfur trioxide is dissolved in
sulfuric acid to form oleum.
This is then reacted with
water to produce more
sulfuric acid.

INDUSTRIAL REACTIONS

Chemicals such as chlorine, oxygen, nitrogen, ammonia, sulfuric acid,
and sodium hydroxide are used in the manufacture of countless
products. Making these chemicals at an industrial scale is an
engineering challenge in which heat, pressure, catalysts, and the
ingenious design of processes and equipment are deployed
to maximize yield and efficiency while keeping costs low.

Targeted delivery
Nanostructures may soon be used in the targeted delivery of drugs to exactly where they are needed in the body of a patient.

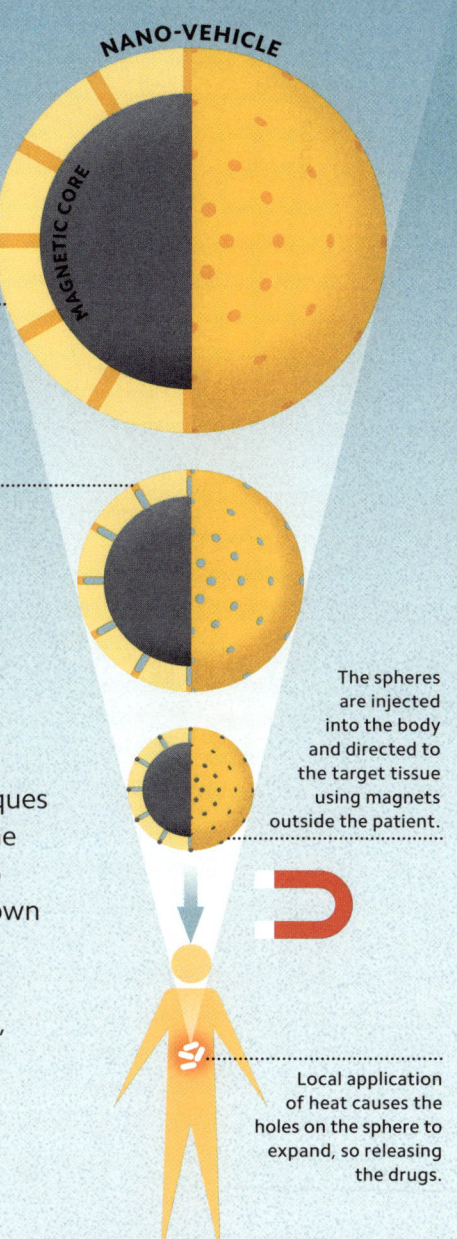

NANO-VEHICLE

MAGNETIC CORE

A tiny sphere of silica is filled with a magnetic core.

Miniature holes in the surface of the sphere are loaded with drug molecules.

SMALL WORLD

While conventional chemistry deals with the bulk properties of materials, the relatively new techniques of nanotechnology ("nano" being the prefix meaning one-billionth) aim to create useful structures at scales down to the size of individual atoms and molecules. Minuscule arrangements of atoms capable of sensing change, manipulating matter, and providing great strength with negligible size have the potential to revolutionize every field of technology – from medicine to agriculture, and computing to construction.

The spheres are injected into the body and directed to the target tissue using magnets outside the patient.

Local application of heat causes the holes on the sphere to expand, so releasing the drugs.

PHYSI

C S

The science of physics deals with the most fundamental of all phenomena – motion, force, mass, energy, space, and time – and attempts to describe how they are related. Early physicists combined philosophical inquiry with experimental techniques; by the 19th century, they believed that they could account for most of the observed Universe. However, their mechanistic theories fell short when they were applied to atomic-scale observations and at velocities approaching light speed. Concepts such as relativity and quantum mechanics emerged in response, but themselves unveiled a host of new questions about our world.

PUSH AND PULL

Force is the amount of push or pull in a given direction. It is measured in units called newtons (N). One newton is about the force you feel when an apple rests on the palm of your hand. In the world around us, many forces are in balance. The downward force exerted by the apple, for example, is balanced by an equal upward force exerted by your hand, so the apple does not move. However, if forces on an object are not balanced (in scale or direction), the object will accelerate (see pp.50–51) or deform.

CONTACT FORCE

A falling apple meets an equal and opposite force when it hits the ground.

Contact and non-contact forces

Forces can be transmitted by physical impact or contact between objects, or – like gravity and magnetism – act at a distance without direct contact.

FORCE AT A DISTANCE

Two suspended apples exert a tiny gravitational force on one another.

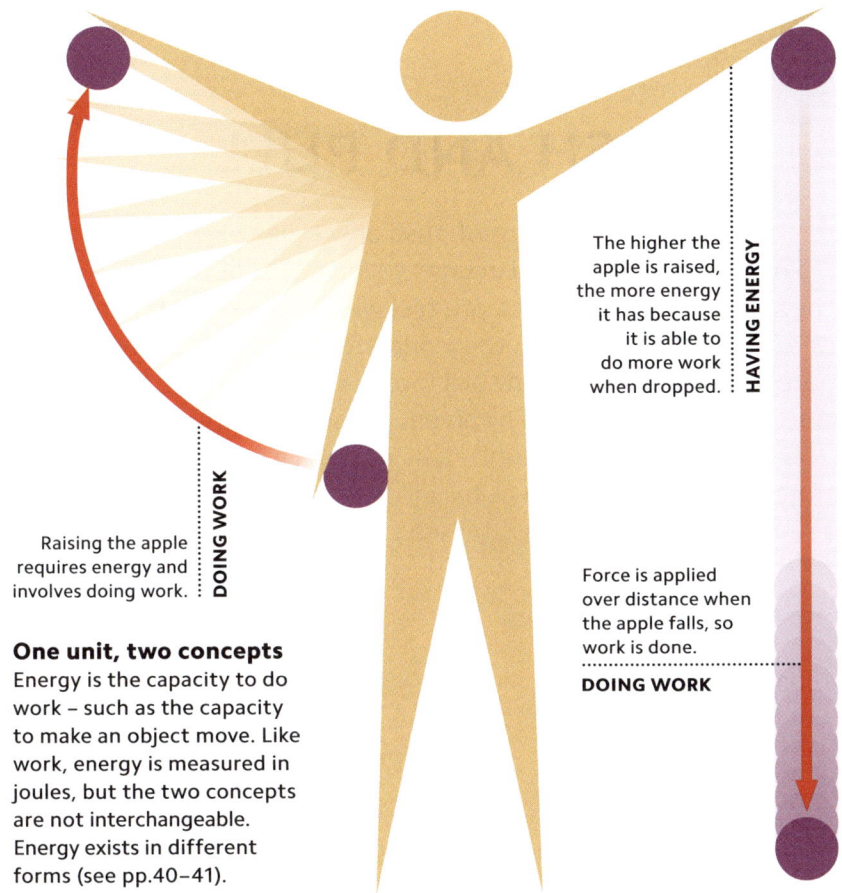

HAVING ENERGY

The higher the apple is raised, the more energy it has because it is able to do more work when dropped.

DOING WORK

Raising the apple requires energy and involves doing work.

DOING WORK

Force is applied over distance when the apple falls, so work is done.

One unit, two concepts

Energy is the capacity to do work – such as the capacity to make an object move. Like work, energy is measured in joules, but the two concepts are not interchangeable. Energy exists in different forms (see pp.40–41).

MAKING A MOVE

The terms work and energy are part of everyday language but have specific meanings to scientists. Work is done when a force moves, or changes the shape of, an object. The force must cause the object to move in the direction of the force: no movement means no work done. The force multiplied by the distance moved (newtons x metres) gives the amount of work done, measured in units called joules (J).

Kinetic energy is proportional to the square of velocity; that is why small, fast objects like bullets carry so much energy.

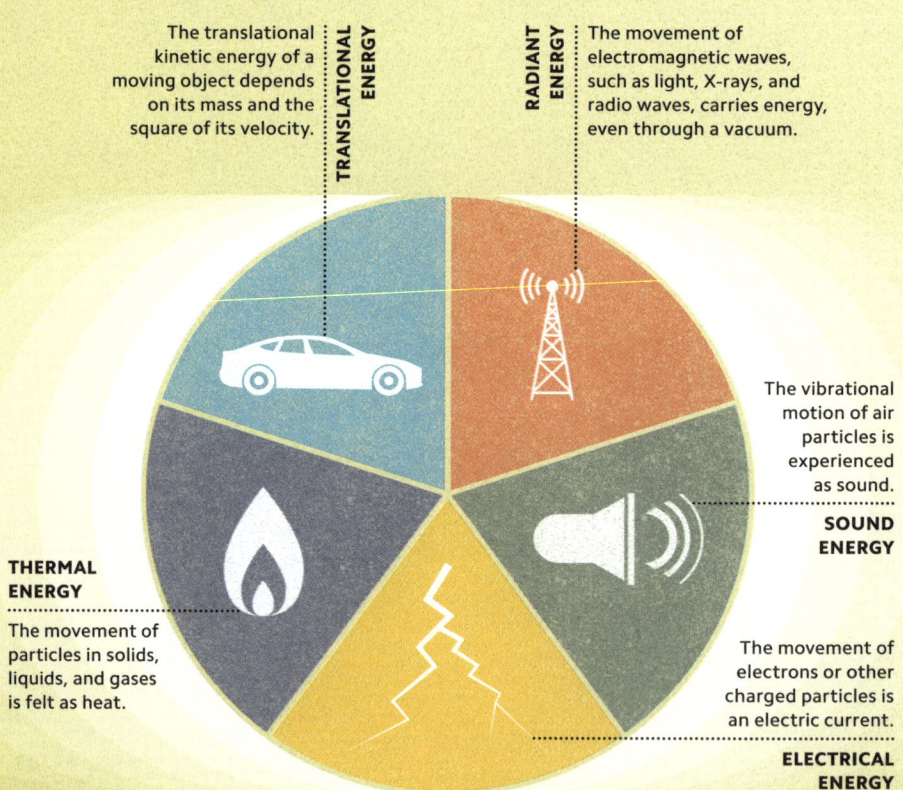

TRANSLATIONAL ENERGY
The translational kinetic energy of a moving object depends on its mass and the square of its velocity.

RADIANT ENERGY
The movement of electromagnetic waves, such as light, X-rays, and radio waves, carries energy, even through a vacuum.

The vibrational motion of air particles is experienced as sound.
SOUND ENERGY

THERMAL ENERGY
The movement of particles in solids, liquids, and gases is felt as heat.

The movement of electrons or other charged particles is an electric current.
ELECTRICAL ENERGY

Kinetic energy
This is the energy possessed by moving things. The greater the velocity and mass of a moving object, the more energy it carries. Kinetic energy takes many different forms.

ENERGETIC WORLD

A moving train clearly has energy, but so does a plank of wood and a tank of water. The wood releases its energy when burned, and the water when it flows. The common thread is that they all have the capacity to do work (see p.39). There are two main types of energy – kinetic and potential – each of which can be seen in different forms. Some forms of energy can be both kinetic and potential.

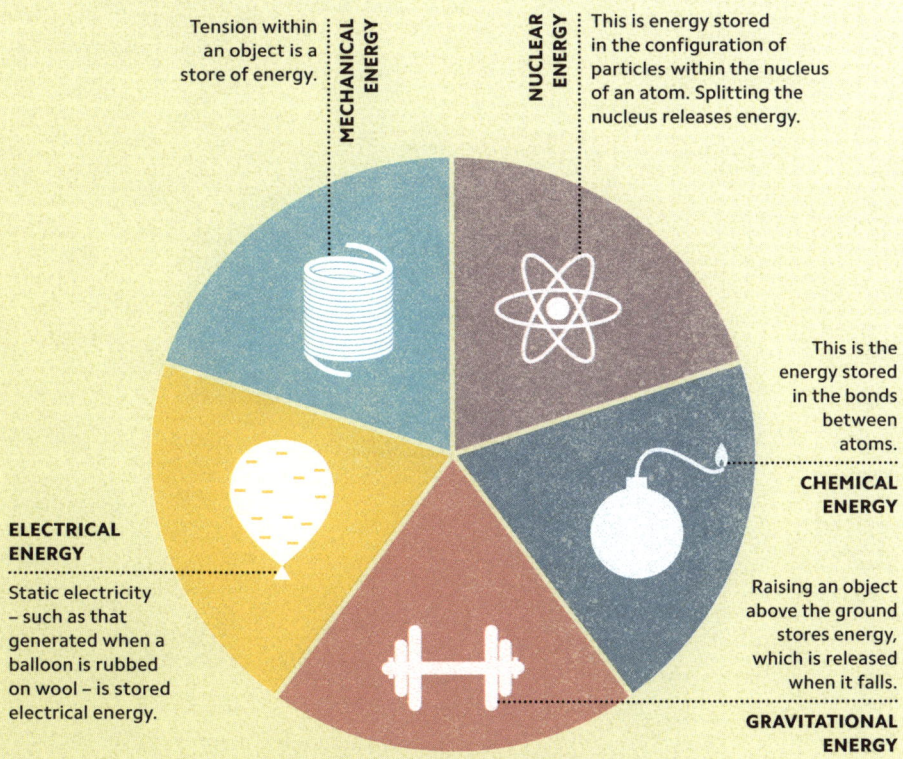

Tension within an object is a store of energy.

MECHANICAL ENERGY

NUCLEAR ENERGY
This is energy stored in the configuration of particles within the nucleus of an atom. Splitting the nucleus releases energy.

This is the energy stored in the bonds between atoms.

CHEMICAL ENERGY

Raising an object above the ground stores energy, which is released when it falls.

GRAVITATIONAL ENERGY

ELECTRICAL ENERGY
Static electricity – such as that generated when a balloon is rubbed on wool – is stored electrical energy.

Potential energy
This is energy stored in an object (which can be a tiny particle) by virtue of its shape or its position relative to other objects. It can take a number of forms.

THE CONSTANCY OF ENERGY

The laws of thermodynamics (see p.46–47) tell us that energy cannot be created from nothing or destroyed. However, it can be changed from one form into another. Whenever we use energy, we convert it into a different type: switching on a lamp, for example, changes electrical energy into radiant energy (light). Such conversions can never be 100 per cent efficient – some energy is always lost as waste heat.

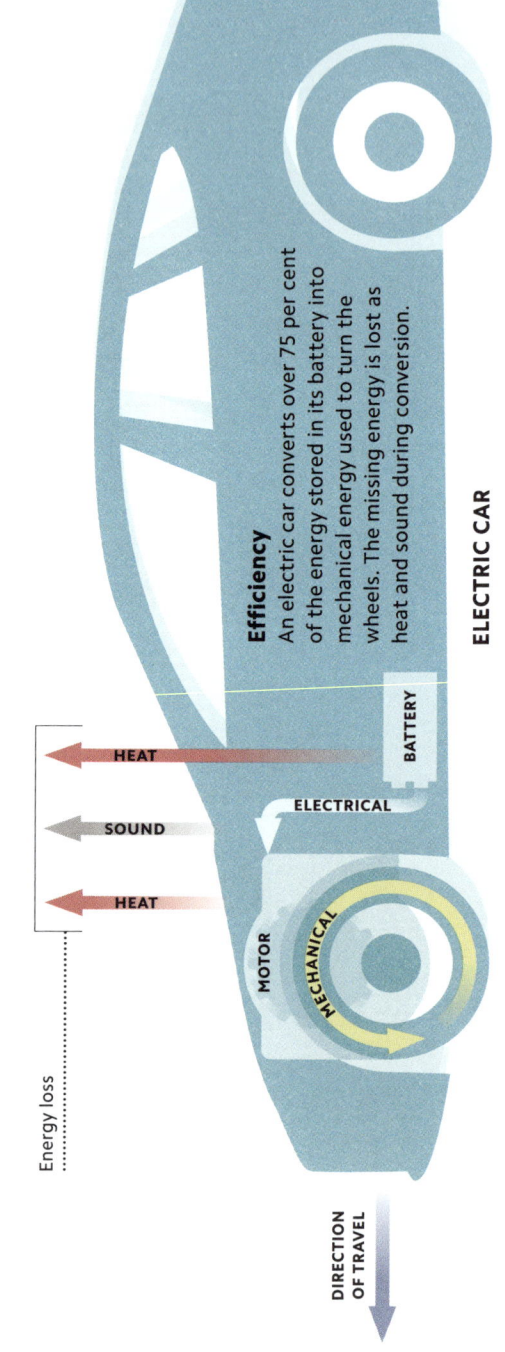

Efficiency

An electric car converts over 75 per cent of the energy stored in its battery into mechanical energy used to turn the wheels. The missing energy is lost as heat and sound during conversion.

ELECTRIC CAR

HEAT

SOUND

HEAT

ELECTRICAL

BATTERY

MOTOR

MECHANICAL

Energy loss

DIRECTION OF TRAVEL

WORK RATE

The scientific definition of power is the rate at which work is done; power is accordingly measured as work (joules) per unit time (second). The unit is the watt (W): one watt equals one joule per second. Power consumption or output is given in watts – so a bulb may consume 10 watts, while an oven may use 3,000 (3 kW).

HORSE
745.7 W

ATHLETE
2,500 W
(2.5 kW)

KETTLE 1,500 W
(1.5 kW)

CAR ENGINE
100,000 W (100 kW)

NUCLEAR POWER
STATION
1,000,000,000 W (1 GW)

Horsepower
Before the watt, power was related to the average rate of work of a dray horse. The peak output of a horse can be more than 15 horsepower.

The power of a rocket at launch is tens of gigawatts.

A HOT TOPIC

Matter is made of particles that are in constant motion. The particles possess energy by virtue of their movement (which includes vibration and rotation) and the forces between them. The temperature of a substance is determined by the average kinetic (movement) energy of its particles. Heat, measured in joules, is the transfer of internal energy from one substance to another, or within a substance. The heat needed to raise the temperature of a body by a set amount depends on the mass of the body and the material of which it is made. For example, heating 1 kg of water by 1 °C (1.8 °F) requires more energy than heating 1 kg of metal by the same amount. Heat can be transferred by conduction, convection, or radiation.

Particles vibrate more slowly

SOLID

TRANSFER OF HEAT

HEAT SOURCE

Particles vibrate rapidly

> "... heat is not a substance, but a dynamical form of mechanical effect..."
> Lord Kelvin

Conduction

In a solid, particles cannot move relative to one another, but they can vibrate. The amount of vibration increases as more heat is supplied. So if one end of a solid bar is heated, the vibrations are passed from one atom to its neighbours, transferring heat along the bar. This is conduction.

HEAT SOURCE

Heat travels through space as waves of infrared radiation.

Cooled liquid descends

LIQUID

TRANSFER OF HEAT

HEAT SOURCE

Heated liquid rises

Particles away from source of radiative heat are eventually warmed by conduction.

Convection

In a fluid (liquid or gas), particles can move relative to one another. Applying heat to one area makes the fluid expand there, so it becomes locally less dense. The less dense region rises, transferring heat within the fluid, before cooling and sinking. This form of heat transfer is convection.

Radiation

Heat can be transferred without contact. The Sun, for example, warms Earth across the vacuum of space. The energy is emitted and travels in the form of electromagnetic radiation (see p.70–71). When radiation is absorbed by a body, it increases the vibration of the body's particles, which raises its temperature.

The four laws

There are four laws of thermodynamics that link temperature, heat, work, and entropy.

Zeroth law

This law concerns the state of thermal equilibrium – when zero heat energy flows between two objects in contact. It states that when two objects are separately in thermal equilibrium with a third object, then they are also in thermal equilibrium with each other.

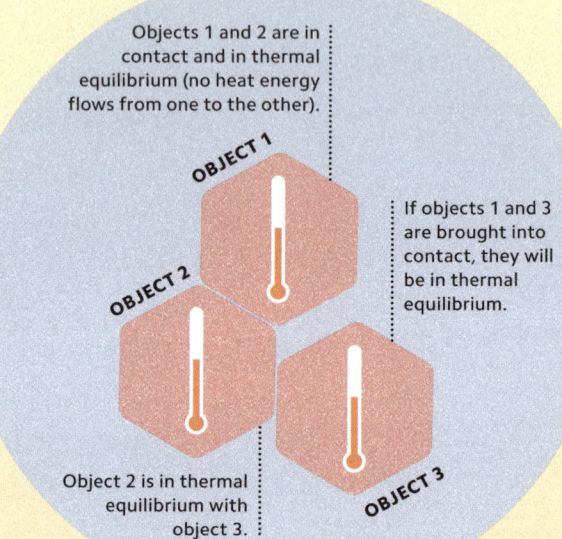

Objects 1 and 2 are in contact and in thermal equilibrium (no heat energy flows from one to the other).

OBJECT 1

OBJECT 2

If objects 1 and 3 are brought into contact, they will be in thermal equilibrium.

Object 2 is in thermal equilibrium with object 3.

OBJECT 3

HEAT, MATTER, AND DISORDER

Thermodynamics deals with the relationships between temperature, heat, and energy. It focuses on internal energy – the energy possessed by atoms and molecules by virtue of their motion and the bonds that hold them together. A key concept is entropy, which is a measure of how spread out the internal energy is within a system. It is also a statistical phenomenon: it is much more likely that energy is spread out than concentrated. Entropy generally increases with temperature: in a solid, for example, particles can only vibrate, but in a liquid, they can vibrate, spin, and move. Entropy within a system never decreases without an input of energy.

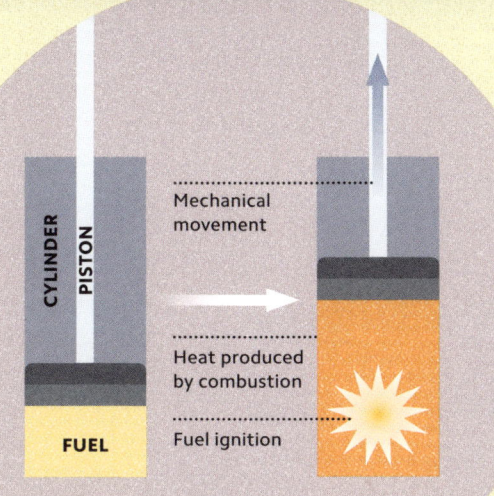

First law

Energy cannot be created or destroyed. It can only be converted into other forms. In a car engine, for example, the chemical energy in the fuel is converted into heat and mechanical movement.

Second law

Heat never spontaneously passes from a colder body to a warmer one. Another way to express this is to state that the degree of disorder – or entropy – of a system will always increase unless more energy is put into the system.

Heat passes spontaneously from a warm mass to a cooler one but never the other way round.

Once the two masses are at thermal equilibrium, the total entropy of the system will have increased.

HOT WATER

COLD WATER

At normal temperatures, particles vibrate and move around.

At absolute zero, molecular movement all but ceases, very few states are available, and systems have zero entropy.

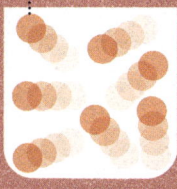

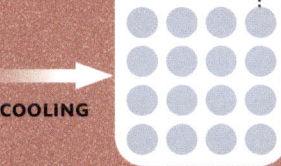

ROOM TEMPERATURE

COOLING

ABSOLUTE ZERO

Third law

Entropy approaches zero near absolute zero (-273 °C or -459 °F) – the temperature at which the vibration of atoms and molecules is at a minimum. This is because, with so little energy, there are very few possible states available to the atoms and molecules.

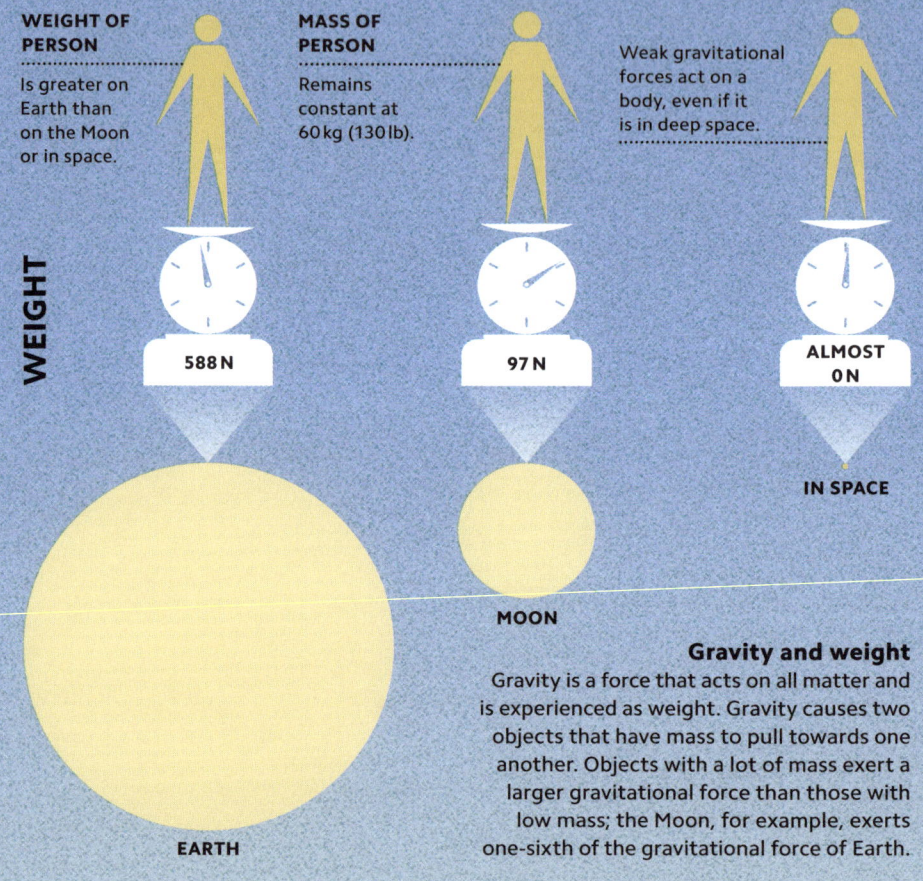

WEIGHT

WEIGHT OF PERSON
Is greater on Earth than on the Moon or in space.

MASS OF PERSON
Remains constant at 60 kg (130 lb).

Weak gravitational forces act on a body, even if it is in deep space.

588 N

97 N

ALMOST 0 N

IN SPACE

EARTH

MOON

Gravity and weight

Gravity is a force that acts on all matter and is experienced as weight. Gravity causes two objects that have mass to pull towards one another. Objects with a lot of mass exert a larger gravitational force than those with low mass; the Moon, for example, exerts one-sixth of the gravitational force of Earth.

WEIGHTY MATTERS

Mass is the amount of matter within an object – the sum of all of its constituent atoms (see pp.8–9). An object's mass, which is measured in kilograms (kg), stays the same wherever that object is located – on Earth, on the Moon, or in outer space. It is sometimes confused for weight, which is a different, though related, concept. Weight is the force of gravity on a mass, and so is measured in newtons (N), as are all forces (see p.38).

EFFORT ▷◁ EFFORT

Amount of force applied

DIRECTING FORCE

The basic definition of a machine is any device that can do work (see p.39). The most useful everyday machines are those that multiply or redirect forces. Levers, as found in a pair of pliers, are simple machines; when you apply a force at one end, a larger force is exerted at the other. A fixed ramp is a machine because pushing a trolley horizontally causes it to rise up, redirecting the force. Similarly, a rope and pulley is a machine because a downward pull on one end of the rope is changed into an upward pull at the other.

WORK

Energy is not created or destroyed by machines. The amount of work done at one end of the lever is equal to that done at the other. The factor by which the machine multiplies force is called its mechanical advantage.

FULCRUM

LEVER LENGTH

Moving the effort further from the fulcrum increases force applied at the jaws, but the effort must be applied over a longer distance.

Mechanical advantage
A pair of pliers is composed of two levers. Applying effort on the long handles produces a larger force at the jaws.

Amount of force exerted

LOAD

> " A change in motion is proportional to the motive force impressed..."
>
> Isaac Newton

VELOCITY 20 KPH / 12 MPH AT 33° INCLINE

UPHILL MOTION

VELOCITY 11 KPH / 7 MPH

Speed and velocity have the same numerical value.

SPEED 20 KPH / 12 MPH

VELOCITY 20 KPH / 12 MPH TO THE RIGHT

VELOCITY 16 KPH / 10 MPH

FORWARD MOTION

Velocity
This is the rate of change of position in a particular direction. It can be measured in metres per second or in kilometres or miles per hour.

Velocity can be resolved into two vectors at right angles to each other.

A stationary car has an acceleration of zero.

As soon as the car starts moving, it is accelerating.

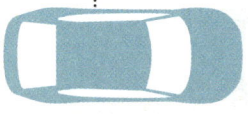

STATIONARY

ACCELERATING

Acceleration
This is the rate of change of velocity. Its scientific units are metres per second, per second.

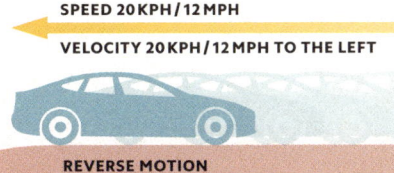

BODIES IN MOTION

The speed of an object is how much distance it covers in a set amount of time – metres per second or miles per hour, for example. A related quantity that is often more useful in science is "velocity", which is an object's speed in a particular direction. Velocity is a vector quantity, which means it has magnitude (size – here, the speed) and direction. Acceleration describes how rapidly the velocity of an object is changing. Like velocity, it is a vector, so a moving object accelerates when it speeds up in a straight line, or if it simply changes direction.

A car driving at constant speed on a circular track is always accelerating because it is constantly changing its direction.

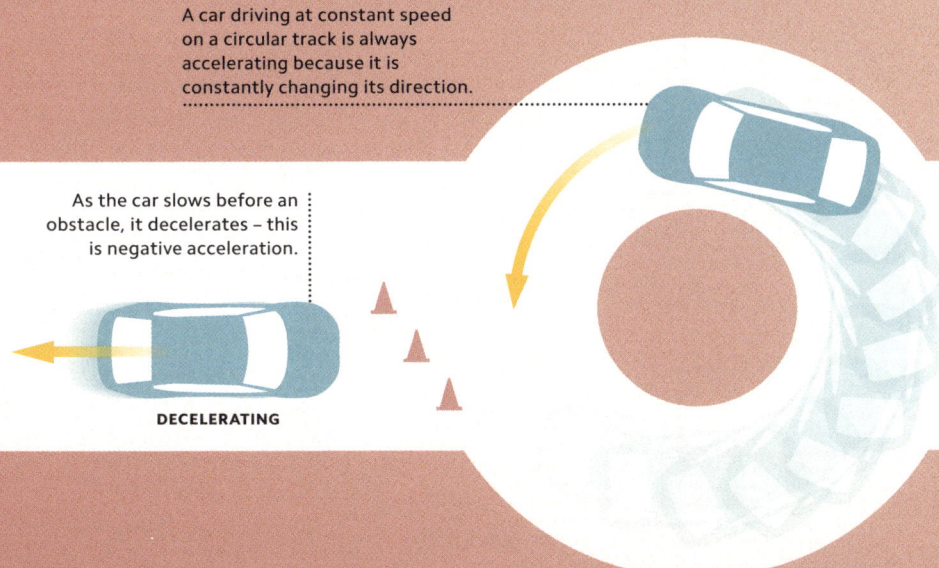

As the car slows before an obstacle, it decelerates – this is negative acceleration.

DECELERATING

NEWTON'S MECHANICS

The relationship between force and motion had been studied by philosophers for centuries, but was given form in the 17th century by the great thinker Isaac Newton. His three laws of motion could be applied to all sorts of systems – from machines to planets and stars – and reshaped our understanding of physics. His laws do, however, break down when objects move close to the speed of light (see p.80).

NEWTON'S SECOND LAW

A force acting on an object will make that object accelerate. The amount of acceleration is proportional to the force.

NEWTON'S THIRD LAW

If object A exerts a force on (pushes against) object B, then object B will push back on object A with the same force in the opposite direction.

The force of the gases expelled from the rocket results in a reactive force pushing the rocket upwards.

NEWTON'S FIRST LAW

An object that is stationary will stay stationary, and one that is moving will keep moving at a constant velocity unless it meets some external force.

MOVING MASSES

When describing how moving objects behave, physicists use concepts including velocity, acceleration (see pp.50–51), momentum, and inertia. The last two on the list seem similar but have different meanings. Momentum is a measure of how much "motion" an object has. It is a multiple of the mass (m) and velocity (v) of that object, so a heavy, fast-moving object has a lot of momentum, whereas a stationary object has none. In contrast, inertia is an innate property of all matter, whether it is moving or not, and depends only on its mass. It is defined as an object's tendency to resist forces that try to change its motion.

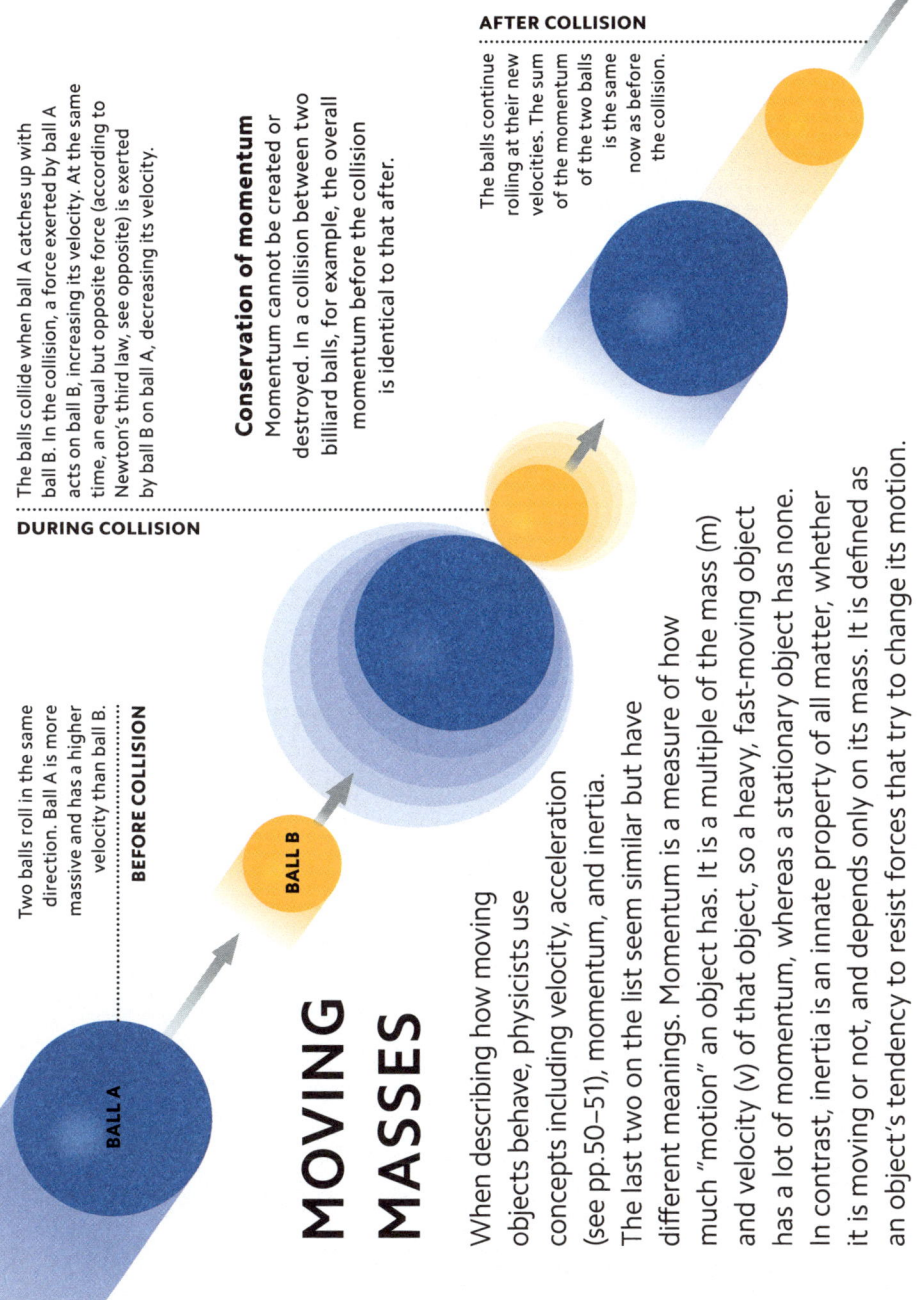

BEFORE COLLISION

Two balls roll in the same direction. Ball A is more massive and has a higher velocity than ball B.

BALL A

BALL B

DURING COLLISION

The balls collide when ball A catches up with ball B. In the collision, a force exerted by ball A acts on ball B, increasing its velocity. At the same time, an equal but opposite force (according to Newton's third law, see opposite) is exerted by ball B on ball A, decreasing its velocity.

Conservation of momentum
Momentum cannot be created or destroyed. In a collision between two billiard balls, for example, the overall momentum before the collision is identical to that after.

AFTER COLLISION

The balls continue rolling at their new velocities. The sum of the momentum of the two balls is the same now as before the collision.

GOING IN CIRCLES

EARTH

A drum of a spinning washing machine and a satellite orbiting Earth are examples of the many objects that move in circles. An object that takes a circular path is constantly changing direction, or accelerating (otherwise it would move in a straight line). Newton's second law tells us that a force is needed to produce acceleration (see p.52), and in the case of circular motion, this force always acts towards the centre of rotation – the axis of the drum or the centre of the Earth. It is called a centripetal (or "centre-seeking") force.

SATELLITE

CENTRIPETAL FORCE

Gravity is the centripetal force that constantly pulls the satellite towards Earth, keeping it on its circular path.

SATELLITE ORBIT

The satellite's velocity is set to counteract the centripetal force; thus it remains in orbit.

VELOCITY

At any one instant, the satellite is moving in a straight line that would take it out of orbit.

The time taken for the pendulum to complete one full cycle of oscillation depends on the length of the pendulum, but, surprisingly, not on its mass.

In SHM, a "restoring force" pulls an object towards its equilibrium position – the point at which it would naturally come to rest. In a pendulum, the restoring force is a combination of gravity and tension.

EQUILIBRIUM POSITION

Maximum restoring force, zero velocity

Minimum restoring force, maximum velocity

TO AND FRO

WAVE MOTION

The movement of the pendulum – or any object in SHM – can be plotted on a graph, where it makes a wave shape.

EQUILIBRIUM POSITION

TIME

A grandfather clock keeps time because its pendulum completes its cycle of motion – from left to right and back again – in the same period, over and over again. This phenomenon – regular movement about a central mid-point – is called simple harmonic motion (SHM), and it applies not only to physical oscillations, such as in timepieces, but also to electromagnetic waves, such as light and radio, sound waves, and the vibration of atoms in crystals.

UNDER PRESSURE

Liquids and gases are fluids – substances that can flow. They have no predefined shape because their molecules can move past one another. There is little space between the molecules of a liquid, so these fluids are difficult (but not impossible) to compress. The molecules of a gas, however, are much more spaced out, so gases can be compressed or expanded by changing either their temperature or pressure. The relationships between their pressure, temperature, and volume are expressed in the gas laws.

The gas laws

The laws that describe the behaviour of gases were set out in the 17th century. They make the (incorrect) assumptions that there is zero force between gas molecules and that the molecules themselves occupy zero volume, so they deviate a little from accurate observations.

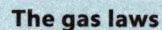
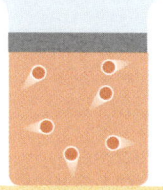
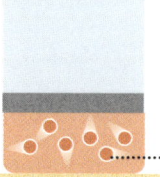
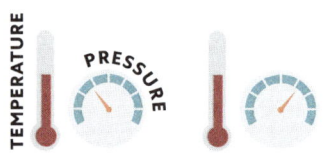

Particles are faster-moving and further apart.

Pressure rises as particles hit the walls more often.

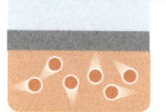
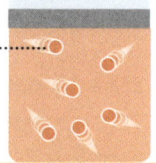

Pressure and volume
The volume of a gas decreases as pressure increases, and vice versa. This relationship is called Boyle's law.

Volume and temperature
The volume of a gas increases as temperature increases. This relationship is called Charles's law.

An object surrounded by a fluid experiences pressure that is exerted by the weight of fluid around it. The pressure depends upon the density of the fluid and increases with depth. If an object is submerged in a fluid, pressure pushing upwards on the bottom of the object is opposed by the pressure pushing downwards on the top. The difference between the two is a buoyant force, called upthrust – and if it is greater than the object's weight, the object will float.

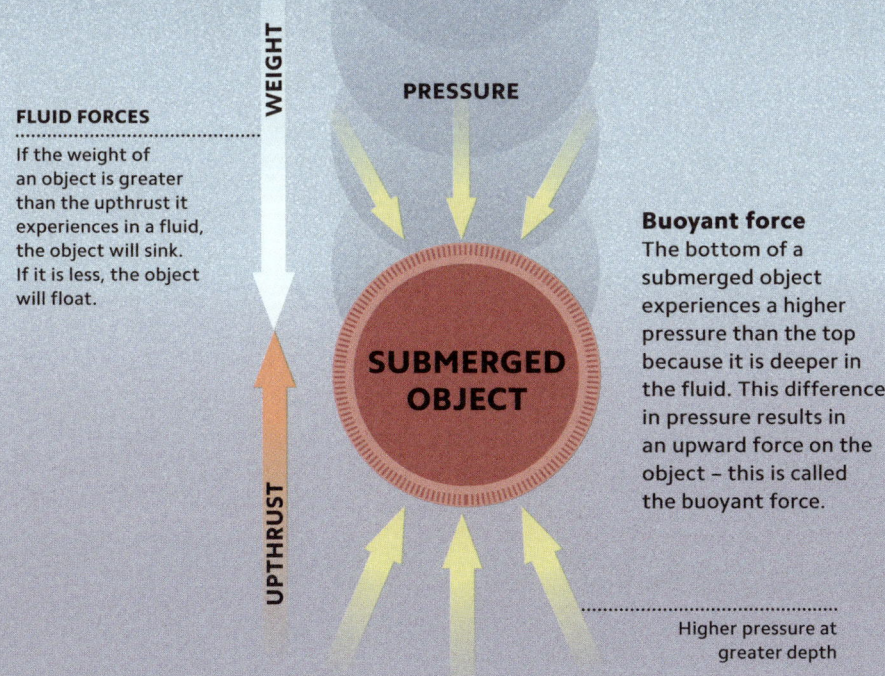

WEIGHT

PRESSURE

FLUID FORCES

If the weight of an object is greater than the upthrust it experiences in a fluid, the object will sink. If it is less, the object will float.

Buoyant force

The bottom of a submerged object experiences a higher pressure than the top because it is deeper in the fluid. This difference in pressure results in an upward force on the object – this is called the buoyant force.

SUBMERGED OBJECT

UPTHRUST

Higher pressure at greater depth

Wave anatomy

Waves can be described by their frequency, wavelength, and amplitude. Waves with a high frequency and amplitude transfer more energy than those with a low frequency and amplitude. The speed of a wave is its frequency multiplied by its wavelength; so the higher the frequency, the smaller the wavelength (assuming that speed remains constant).

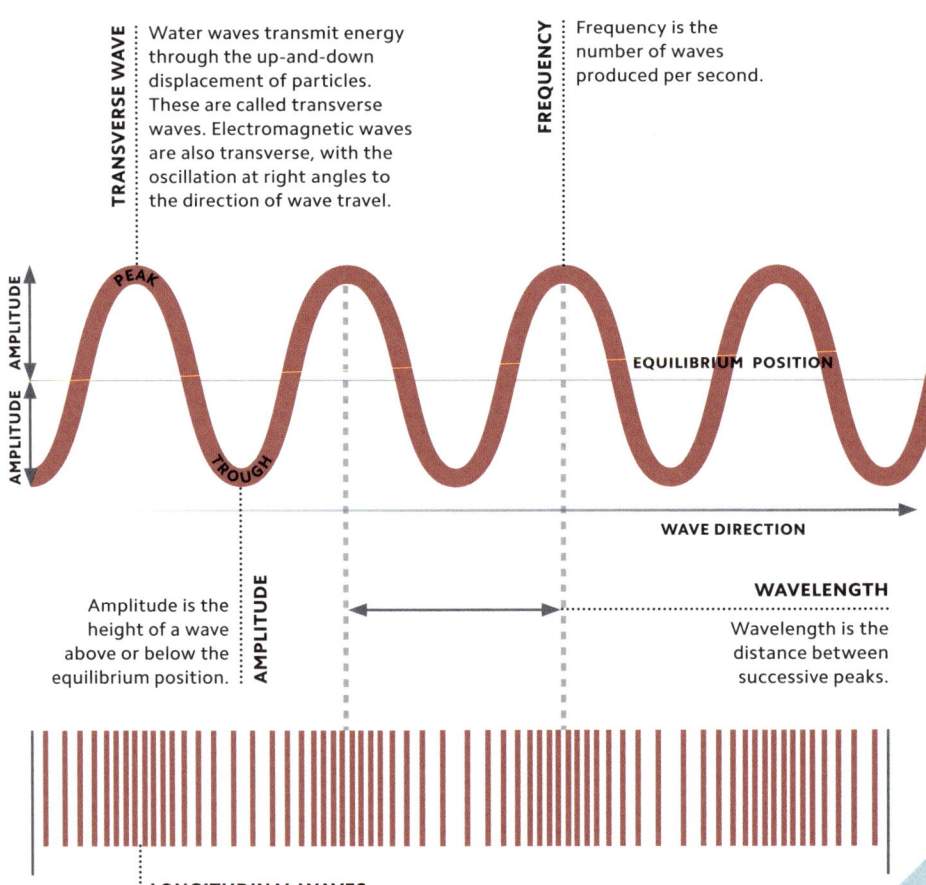

TRANSVERSE WAVE
Water waves transmit energy through the up-and-down displacement of particles. These are called transverse waves. Electromagnetic waves are also transverse, with the oscillation at right angles to the direction of wave travel.

FREQUENCY
Frequency is the number of waves produced per second.

AMPLITUDE

AMPLITUDE

PEAK

TROUGH

EQUILIBRIUM POSITION

WAVE DIRECTION

AMPLITUDE
Amplitude is the height of a wave above or below the equilibrium position.

WAVELENGTH
Wavelength is the distance between successive peaks.

LONGITUDINAL WAVES
Sound waves, some earthquake waves, and slinky springs transmit energy by compressing and then stretching out the medium that they move through. These are called longitudinal waves. They oscillate in the direction of wave travel.

REFLECTED WAVE
....
All waves can be reflected when
they arrive at a fixed boundary
– that is why you will hear an
echo in a mountain valley and
see a reflection in a mirror.

WAVE DIRECTION

AIR

GLASS

WAVE DIRECTION

Ripples on
a pond, radio
signals, sound,
and earthquakes
appear to have little in
common, but they are all
types of wave – a travelling
disturbance that transmits
energy from place to place.
There are two main types of wave:
mechanical and electromagnetic. In a
mechanical wave, the disturbance is
physical, while in an electromagnetic
wave (see pp.70–71) it is made up of
electric and magnetic fields.
Mechanical waves require a
physical medium through
which to move;
electromagnetic
waves do not.

The path of a wave
may be deflected (or
refracted) when it
passes into a medium
of different density –
light passing from air
into glass, for example.

REFRACTED WAVE

PULSING PRESSURE

Sound is a form of mechanical wave – an oscillation within a substance. Unlike a water wave, where the particles of water move at right angles to the direction of wave motion, sound is a lengthways, or longitudinal, oscillation, in which vibrations occur along the same direction as the wave's motion. Like all mechanical waves, sound is transmitted through particles – it cannot travel through a vacuum. In air, sound travels at about 1,200 kph (750 mph), but moves more quickly through liquids and solids – materials made of more closely packed particles; its speed through metal, for example, is about 18,000 kph (11,000 mph).

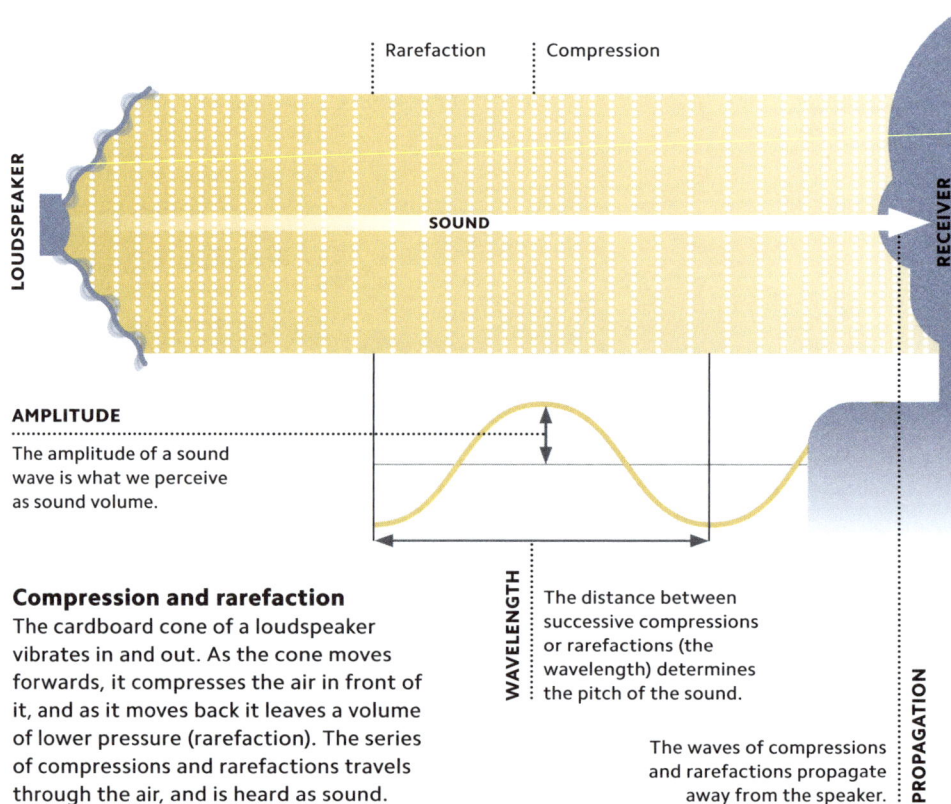

Rarefaction

Compression

LOUDSPEAKER

SOUND

RECEIVER

AMPLITUDE
The amplitude of a sound wave is what we perceive as sound volume.

WAVELENGTH : The distance between successive compressions or rarefactions (the wavelength) determines the pitch of the sound.

Compression and rarefaction
The cardboard cone of a loudspeaker vibrates in and out. As the cone moves forwards, it compresses the air in front of it, and as it moves back it leaves a volume of lower pressure (rarefaction). The series of compressions and rarefactions travels through the air, and is heard as sound.

PROPAGATION : The waves of compressions and rarefactions propagate away from the speaker.

VISIBLE EVIDENCE

Waves on water and some other mechanical waves can be observed directly, but why is light considered to be a wave? The evidence comes from studies of a phenomenon called diffraction, in which waves spread out as they move past edges or through gaps. Diffraction happens with all types of waves, and was first observed to occur with light in the 16th century, indicating that light too has wavelike properties.

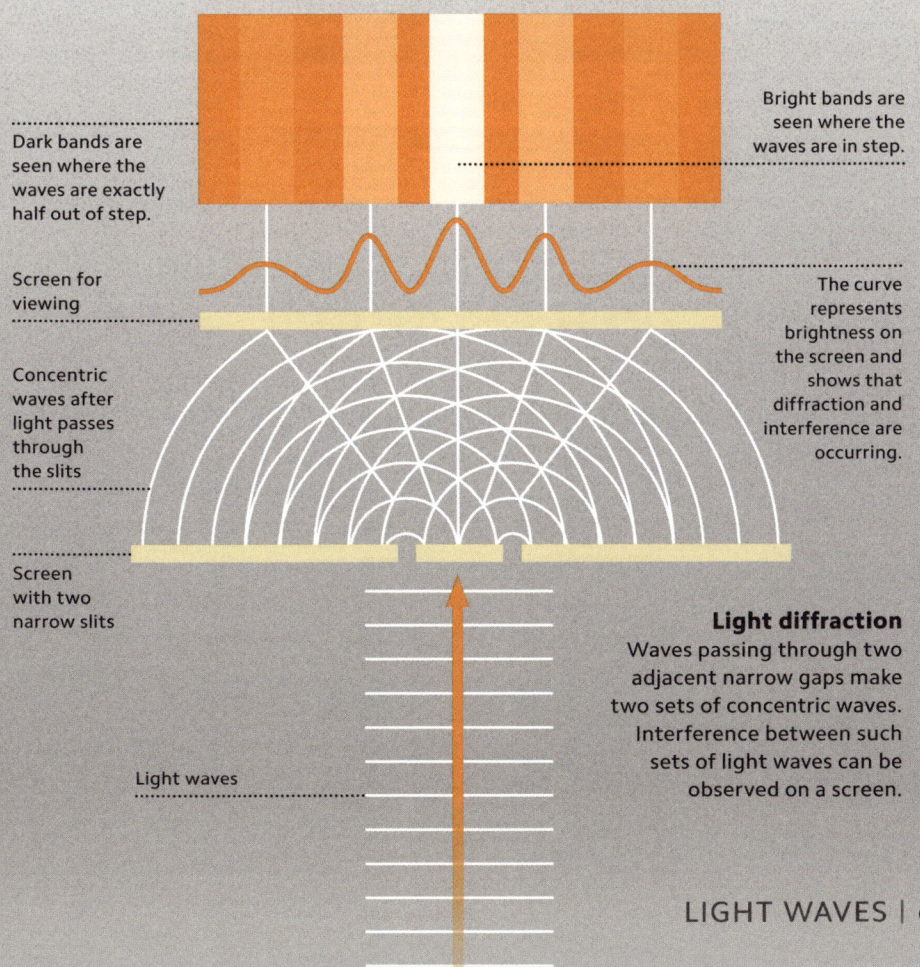

Bright bands are seen where the waves are in step.

Dark bands are seen where the waves are exactly half out of step.

The curve represents brightness on the screen and shows that diffraction and interference are occurring.

Screen for viewing

Concentric waves after light passes through the slits

Screen with two narrow slits

Light diffraction
Waves passing through two adjacent narrow gaps make two sets of concentric waves. Interference between such sets of light waves can be observed on a screen.

Light waves

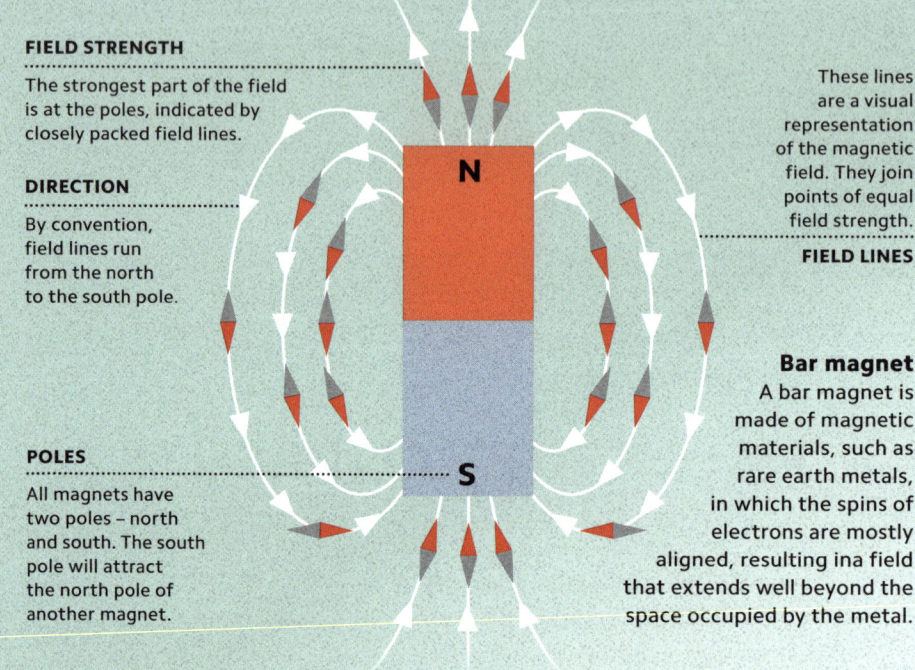

FIELD STRENGTH

The strongest part of the field is at the poles, indicated by closely packed field lines.

DIRECTION

By convention, field lines run from the north to the south pole.

POLES

All magnets have two poles – north and south. The south pole will attract the north pole of another magnet.

These lines are a visual representation of the magnetic field. They join points of equal field strength.

FIELD LINES

Bar magnet

A bar magnet is made of magnetic materials, such as rare earth metals, in which the spins of electrons are mostly aligned, resulting in a field that extends well beyond the space occupied by the metal.

POLES APART

Magnetism is a force that acts on objects at a distance, without the need for physical contact. The area around a magnetic object in which this force is felt is called a magnetic field. It is created by the movement of charged particles, such as electrons and ions. Some charged particles, such as electrons, have a property called spin – the equivalent of them rotating on their axes. This motion produces a magnetic field. In most materials the particles' spins are randomly oriented, so there is no overall magnetic field. In magnetic materials, such as iron and nickel, the spins can be made to align, resulting in a strong overall field in a solid bar of the metal.

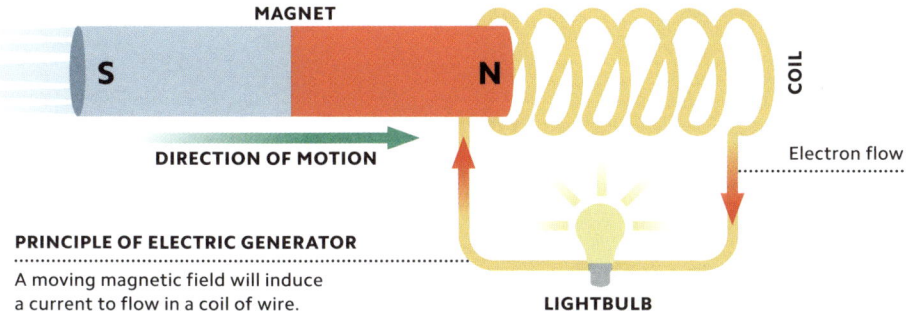

MAGNET

S

N

COIL

DIRECTION OF MOTION

Electron flow

PRINCIPLE OF ELECTRIC GENERATOR

A moving magnetic field will induce
a current to flow in a coil of wire.

LIGHTBULB

FIELDS AND CURRENTS

Just as a spinning electron in an atom creates a magnetic field (see opposite), so does a stream of electrons (an electric current) moving through a conductor, such as a copper wire. Conversely a moving magnet causes electrons to flow in a conductor that is within the magnetic field. These complementary facts reveal that electricity and magnetism are two aspects of a single phenomenon called electromagnetism. This fundamental force not only makes electric motors and generators work, but it underlies the nature of electromagnetic waves, including light and radio waves (see pp.70–71).

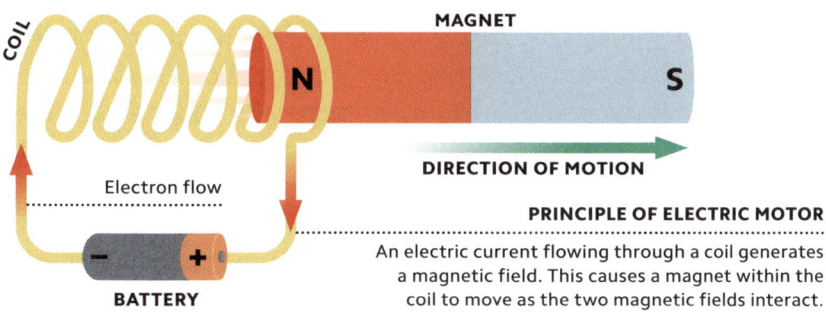

COIL

N

MAGNET

S

Electron flow

DIRECTION OF MOTION

PRINCIPLE OF ELECTRIC MOTOR

An electric current flowing through a coil generates
a magnetic field. This causes a magnet within the
coil to move as the two magnetic fields interact.

BATTERY

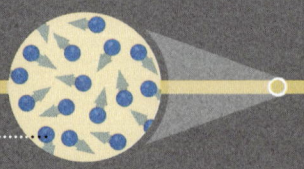

FREE ELECTRONS

Electrons in a wire not
connected to a battery
move freely at random.

COPPER WIRE

Copper is a good
electrical conductor.

THE MOVE

CHARGE ON

Current is the rate of flow of charged particles
through a medium. The particles can be ions (see
p.18) or protons, but in many common cases are
electrons. Media in which these particles can flow
are called conductors; ones that prevent flow are
insulators. Metals, such as copper, are conductors
because they contain trillions of electrons that
are only loosely bound to their atoms. Connecting
the terminals of a battery with a wire causes the
electrons to flow. This is a direct electric current.

FLOW OF ELECTRONS

DIRECT CURRENT

A battery converts
chemical energy into
an electric force that
pushes electrons
one way around a
conductive circuit.

Alternating current

A battery (or cell) pushes electrons one way
around a circuit but generators, such as those
that supply power grids used by homes, work by
moving the electrons to and fro. Such an alternating
current is easier to transmit over long distances.

Electrical wall
socket

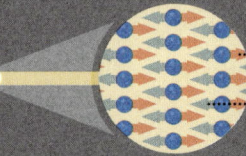

Electrons move
forwards

Electrons move
back

CHARGED UP

An atom is made up of a nucleus, which carries a positive charge, surrounded by a cloud of electrons, each of which has a negative charge. A normal atom carries no net charge. However, electrons can sometimes be gained or lost: even friction can scrape electrons from one atom onto another. This is what occurs when a balloon is rubbed against a woollen jumper, or when air molecules in clouds rub against one another in turbulent air. This imbalance of charges is called static electricity. It can cause objects to attract one another. A release of charge (or discharge) can produce a flash, such as lightning.

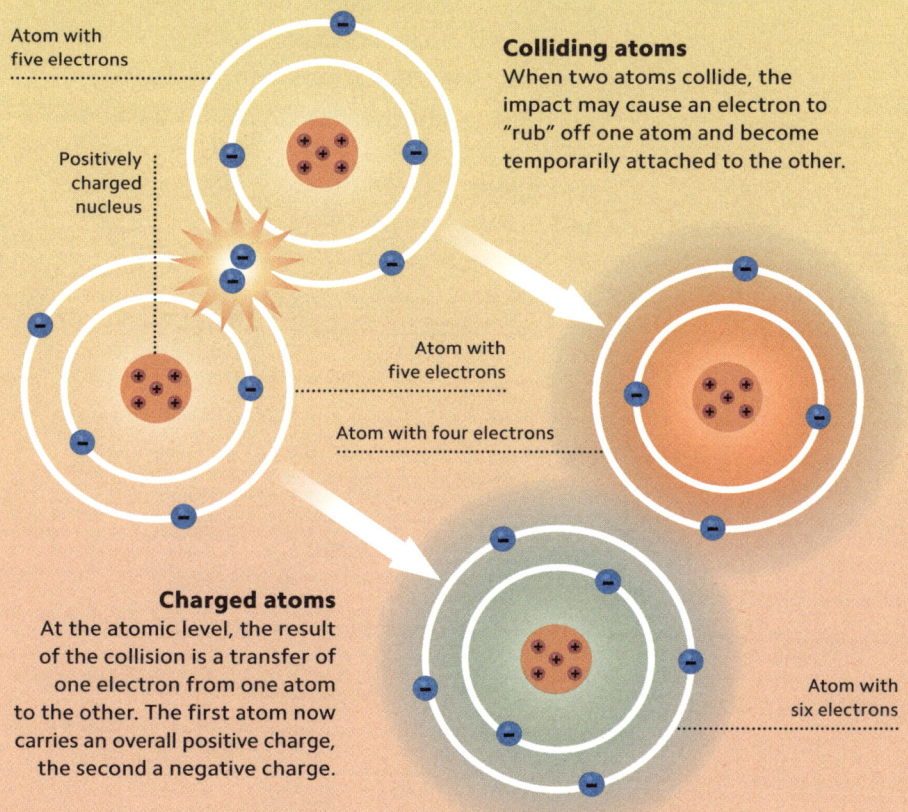

Atom with five electrons

Positively charged nucleus

Colliding atoms
When two atoms collide, the impact may cause an electron to "rub" off one atom and become temporarily attached to the other.

Atom with five electrons

Atom with four electrons

Charged atoms
At the atomic level, the result of the collision is a transfer of one electron from one atom to the other. The first atom now carries an overall positive charge, the second a negative charge.

Atom with six electrons

POWER SUPPLY

A battery or a source of alternating current drives electrons round a circuit. Electrons move from the negative to positive battery terminal. The electromotive force driving the electrons is measured in volts (V).

CONDUCTOR

This carries electrons round the circuit and has a low resistance. The amount of electrons flowing past a point in the circuit in a given time is called the current, and is measured in amps (A).

ROUND AND ROUND

Circuits are central to the operation of all electrical devices. As their name suggests, they must be continuous (or closed) for electric current to flow. Every circuit has three basic parts: a power supply that pushes electrons around the circuit; a conductor (usually copper wire) that carries the electrons; and a load, which has some resistance and performs a particular function. The load can be a light bulb or an array of components such as resistors, diodes, transistors, or inductors that carry out a range of functions, such as amplification or logical operations, depending on how they are configured. Complex equipment is usually built up from multiple circuits, which may each have multiple branches.

CIRCUIT SYMBOLS

Engineers use standard symbols to denote different components in a circuit. This symbol denotes a light bulb.

LOAD

The load resists the flow of electrons; this resistance produces heat, which is why devices get warm with use. Resistance is measured in ohms (Ω).

DOPED CRYSTALS

Crystals of materials such as silicon are known as semiconductors because they behave as electrical insulators under normal conditions but will conduct electricity if heated. Incorporating atoms of other elements into their crystals (known as doping) also changes their properties. These new properties can be harnessed to make components such as diodes and transistors.

Added impurities

A silicon atom has four electrons in its valence shell. In a crystal of silicon, each atom makes four covalent bonds with its neighbours (see p.19). With all its electrons locked up in bonds, silicon behaves as an insulator, but adding atoms of phosphorus or boron changes silicon into a conductor.

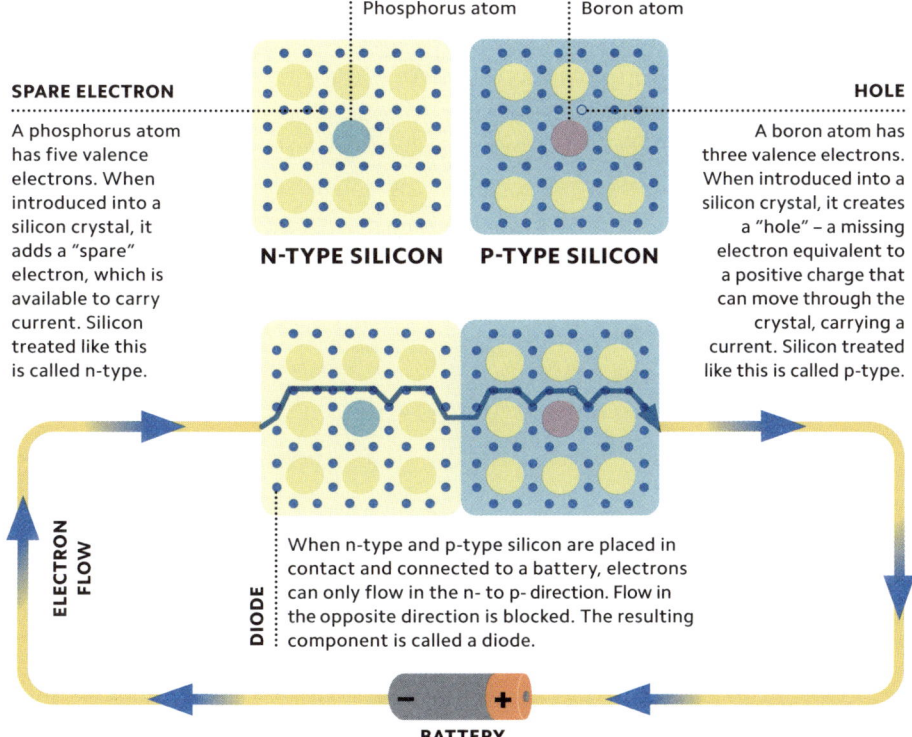

Phosphorus atom

Boron atom

SPARE ELECTRON

A phosphorus atom has five valence electrons. When introduced into a silicon crystal, it adds a "spare" electron, which is available to carry current. Silicon treated like this is called n-type.

HOLE

A boron atom has three valence electrons. When introduced into a silicon crystal, it creates a "hole" – a missing electron equivalent to a positive charge that can move through the crystal, carrying a current. Silicon treated like this is called p-type.

N-TYPE SILICON   **P-TYPE SILICON**

ELECTRON FLOW

DIODE

When n-type and p-type silicon are placed in contact and connected to a battery, electrons can only flow in the n- to p- direction. Flow in the opposite direction is blocked. The resulting component is called a diode.

BATTERY

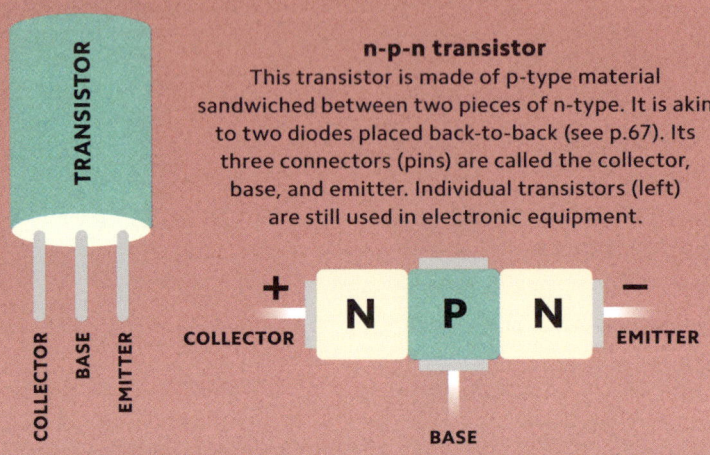

n-p-n transistor

This transistor is made of p-type material sandwiched between two pieces of n-type. It is akin to two diodes placed back-to-back (see p.67). Its three connectors (pins) are called the collector, base, and emitter. Individual transistors (left) are still used in electronic equipment.

TRANSISTOR

COLLECTOR BASE EMITTER

+ **N P N** −

COLLECTOR EMITTER

BASE

MINIATURE SWITCHES

Today's electronic systems rely on semiconductor devices called transistors. They are used to amplify a signal or to switch an electrical current between two terminals on or off, depending on the voltage at the third terminal. Circuits containing many transistors linked in specific configurations can carry out simple logical functions, such as addition and subtraction, or comparing one input with another. Combined, miniaturized, and etched onto silicon chips in their millions, such circuits are the powerhouses behind digital computers.

The transistor as a switch

If a battery is connected between the collector and emitter, no current will flow. However, if a small voltage is applied between base and emitter, the transistor is switched on – a larger current can now flow between collector and emitter.

BASE

COLLECTOR

EMITTER

TRIGGER
A small current switches on the transistor.

Once the transistor is switched on, a larger current can flow from collector to emitter.

CURRENT FLOW

Narrow AI

This is capable of surpassing human intelligence for a single, limited purpose, such as facial recognition, identifying cancerous cells in a tissue sample, playing chess, or serving as a chatbot.

General AI

This AI can use previously gained data and problem-solving abilities in a new environment, without the need for fresh programming. Its learning and problem-solving capabilities match those of a human. It has not yet been achieved.

Super AI

AI in this category has cognitive skills that far surpass human ability. It possesses intentions and emotions of its own, and is capable of abstraction beyond the limits of the human mind. It has not yet been achieved.

BEYOND THE HUMAN MIND

The number of transistors that can be etched onto a single integrated circuit has doubled every two years since the 1970s – a trend known as Moore's Law – making digital computers ever faster and more affordable. However, computers remain somewhat limited because they work by executing sets of preset instructions. Recent advances allow machines to work more like the human mind – adapting and learning to make better decisions. Forms of such artificial intelligence (AI) are today capable of extracting and applying principles by examining existing data, but the potential of AI is far greater. The emergence of a truly autonomous AI may deliver great benefits to humanity, or imperil its existence.

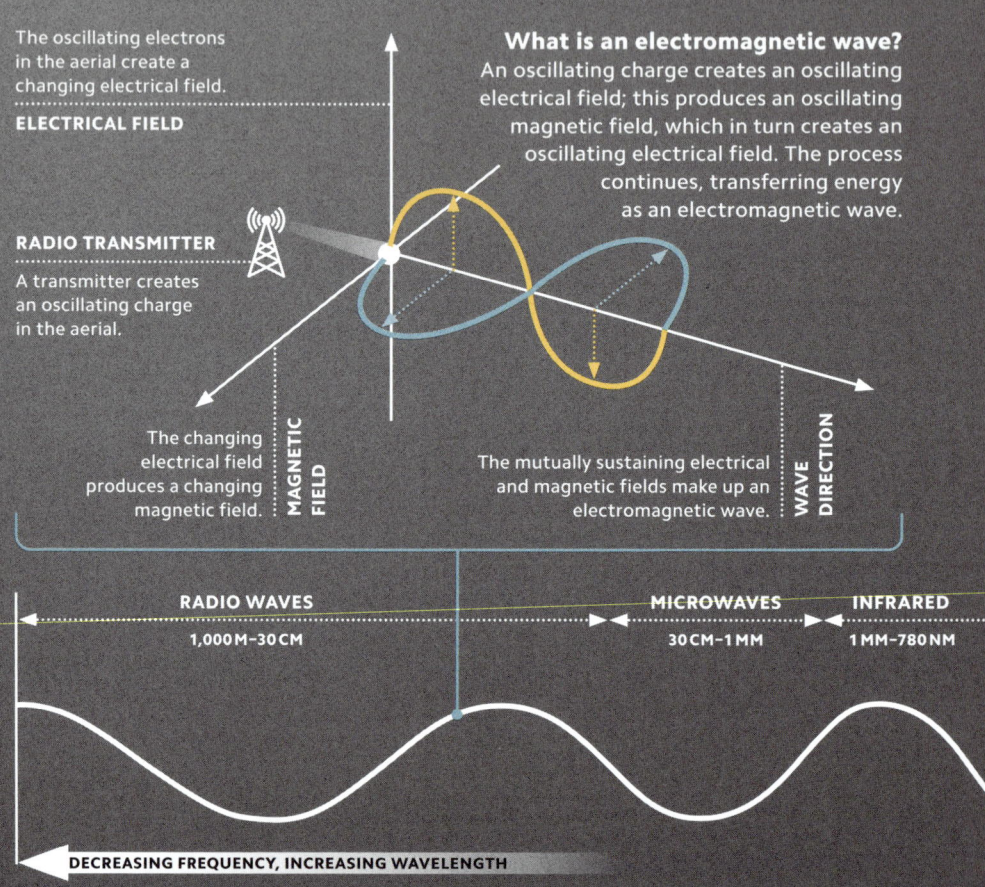

The oscillating electrons in the aerial create a changing electrical field.

ELECTRICAL FIELD

What is an electromagnetic wave?
An oscillating charge creates an oscillating electrical field; this produces an oscillating magnetic field, which in turn creates an oscillating electrical field. The process continues, transferring energy as an electromagnetic wave.

RADIO TRANSMITTER

A transmitter creates an oscillating charge in the aerial.

The changing electrical field produces a changing magnetic field.

MAGNETIC FIELD

The mutually sustaining electrical and magnetic fields make up an electromagnetic wave.

WAVE DIRECTION

RADIO WAVES
1,000M–30CM

MICROWAVES
30CM–1MM

INFRARED
1MM–780NM

DECREASING FREQUENCY, INCREASING WAVELENGTH

The spectrum
The different frequencies of EM radiation can be visualized as a spectrum. Only a small part of this spectrum corresponds to visible light. The higher the frequency of the waves, the higher their energy; X-rays and gamma rays have enough energy to penetrate and damage living cells.

The wavelength of visible light is about the same length as a virus.

SURFING THE SPECTRUM

Mechanical waves, such as sound vibrations and water waves, have a range of frequencies and corresponding wavelengths (see pp.58–59). So too do electromagnetic (EM) waves, which are formed by oscillating electrical and magnetic fields. Light, radio waves, microwaves, X-rays, and gamma rays are all forms of electromagnetic radiation that are defined by their frequency (and thus wavelength). They all travel at a speed of 300,000,000 m (980,000,000 ft) per second in a vacuum, and can be reflected and refracted (bent) when they pass from a medium of one density into another.

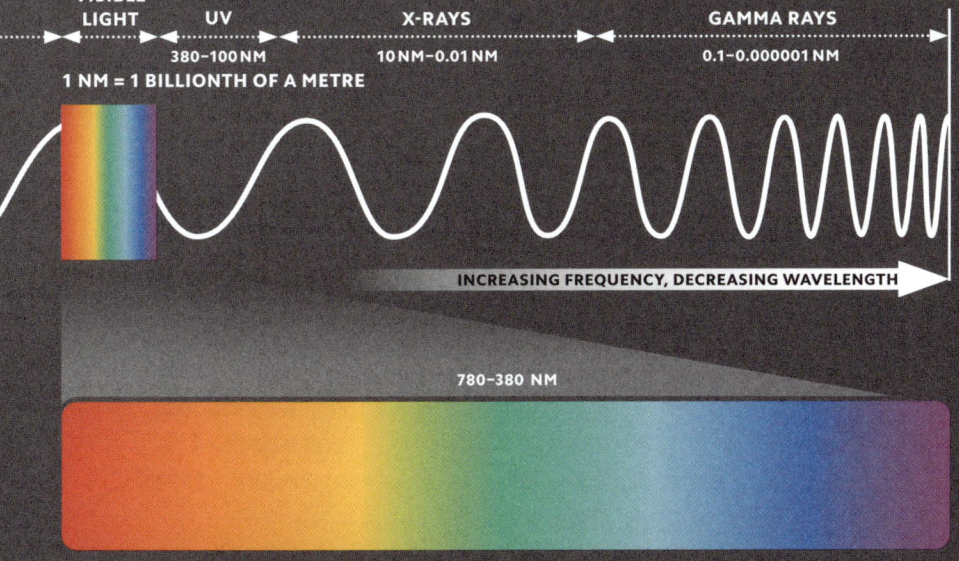

VISIBLE LIGHT

UV
380–100 NM

X-RAYS
10 NM–0.01 NM

GAMMA RAYS
0.1–0.000001 NM

1 NM = 1 BILLIONTH OF A METRE

INCREASING FREQUENCY, DECREASING WAVELENGTH

780–380 NM

Visible light
Colour is not an inherent property of EM radiation; the human brain perceives different wavelengths of light as different colours.

FORCES OF NATURE

The behaviour of all matter is governed by interactions between the elementary particles of which it is made (see pp.74–75). There are four fundamental interactions, or forces: the strong interaction, the weak interaction, the electromagnetic interaction, and gravitation. The first three are described by quantum theories (see pp.76–79) and are carried, or "mediated", by the exchange of fleeting "virtual" particles called gauge bosons. Gravitation has not yet been described by a quantum theory – it is best accounted for by general relativity (see p.81) – but a gauge boson, the graviton, may yet be discovered.

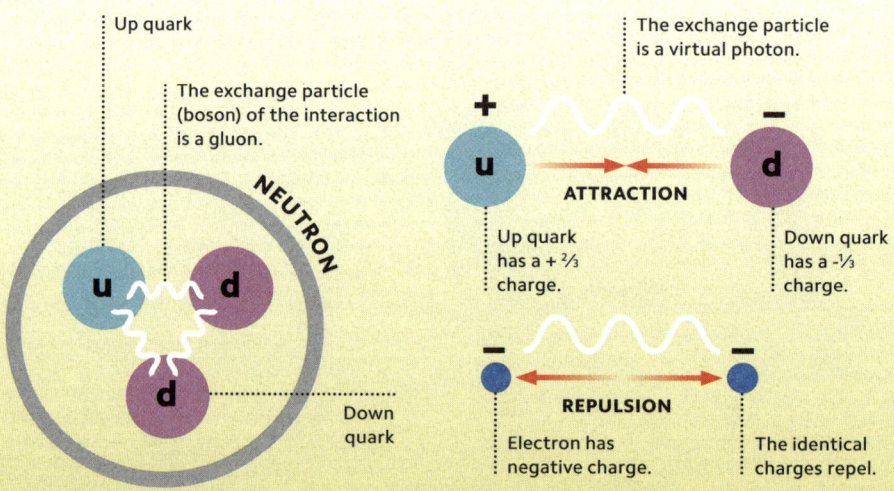

Strong interaction
This force works at very short ranges – around the diameter of a medium-size atomic nucleus. It acts on quarks, holding them together to form hadrons (such as protons and neutrons).

Electromagnetic interaction
This force affects any charged particles and can be attractive or repulsive. It is responsible for holding atoms and molecules together. It has about 0.007 times the strength of the strong interaction.

> The strong interaction is the strongest force in the Universe – 100 trillion trillion trillion times stronger than gravity.

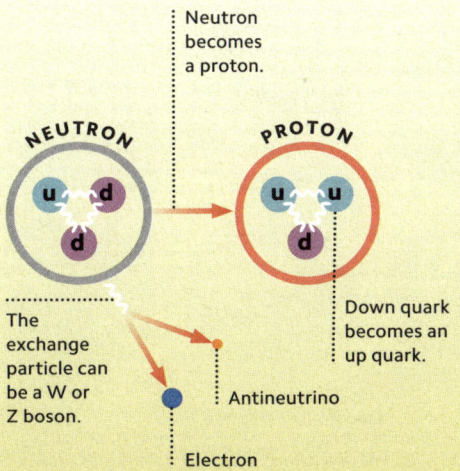

Neutron becomes a proton.

NEUTRON

u d
d

PROTON

u u
d

The exchange particle can be a W or Z boson.

Down quark becomes an up quark.

Antineutrino

Electron

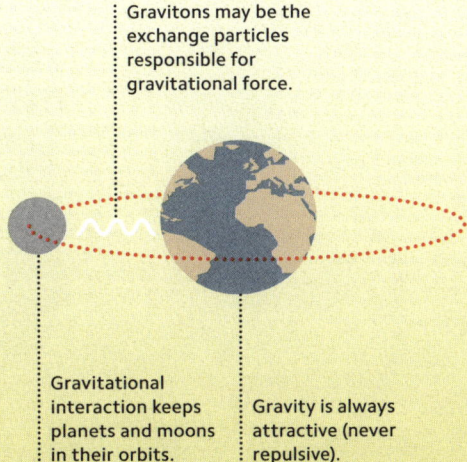

Gravitons may be the exchange particles responsible for gravitational force.

Gravitational interaction keeps planets and moons in their orbits.

Gravity is always attractive (never repulsive).

Weak interaction

This force is about one million times weaker than the strong interaction. It works at very short ranges only – less than the diameter of one proton. It is involved in the decay of certain subatomic particles, such as the decay of neutrons to protons.

Gravitation

This is responsible for the attraction between objects. It has a minute fraction of the strength of the strong interaction but works over vast distances. No quantum theory of gravitation exists to date.

Elementary fermions

These particles include electrons and up and down quarks, which account for almost all visible matter. They spin, and so have momentum, which always has a half-integer value, such as ½ or ³⁄₂.

FERMIONS

QUARKS

u UP	**c** CHARM	**t** TOP

QUARKS

There are six different flavours of quark that differ in mass and electric charge. All are affected by all four fundamental interactions.

d DOWN	**s** STRANGE	**b** BOTTOM

LEPTONS

There are two classes of leptons, charged and uncharged. The most familiar is the (negatively charged) electron. Leptons are affected by all the fundamental interactions except the strong interaction.

LEPTONS

e ELECTRON	**μ** MUON	**τ** TAU
υₑ ELECTRON NEUTRINO	**υμ** MUON NEUTRINO	**υτ** TAU NEUTRINO

INCREASING MASS →

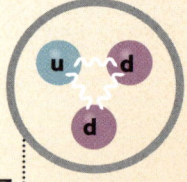
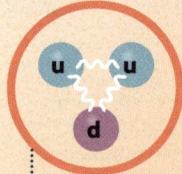

Hadrons

Quarks always combine to form larger particles called hadrons. The most familiar and most stable of these are protons and neutrons.

NEUTRON: Neutrons are composed of one up and two down quarks.

PROTON: Protons are composed of two up and one down quark.

These force carriers are responsible for the electromagnetic interaction.

BOSONS

γ
PHOTON

H
HIGGS BOSON

W
W BOSON

HIGGS BOSON

This particle is a product of the Higgs field, which gives elementary particles their mass.

Z
Z BOSON

W+, W-, AND Z0 BOSONS

These force carriers are responsible for the weak interaction.

g
GLUON

GLUON

These force carriers are responsible for the strong interaction.

GAUGE BOSONS

Elementary bosons

Bosons are particles whose spin momentum number is always a whole number. Four of the elementary bosons are gauge bosons; the other one gives elementary particles their mass.

SMALL WORLD

Matter is ultimately composed of elementary, or fundamental, particles – those that cannot be further divided into smaller components. The current theory that describes particles and their interactions is called the standard model. This model recognizes just two main kinds of elementary particle: elementary fermions and elementary bosons. Fermions take up space, so are considered to be matter; bosons do not take up space, but are responsible for the fundamental forces, or "interactions", of matter (see pp.72–73).

Every fermion has an anti-particle that carries the opposite charge, so there are 24 fermions in all.

STEPPING UP

Just as Newton's laws (see p.52) predict the behaviour of objects at ordinary scales, quantum mechanics deals with matter and energy at the atomic and subatomic levels. It holds that quantities – such as energy and momentum – are not continuously variable but have discrete levels, or "quantized" values, and that tiny particles such as photons and electrons can behave as matter or waves (but not both at the same time). Where classical physics deals with certainty, quantum mechanics deals with probabilities. It predicts the behaviour of particles very well mathematically, but offers up ideas that can appear paradoxical, such as that particles can become entangled, so that their characteristics are interdependent even if they are separated by vast distances.

"How wonderful that we have met with a paradox."

Niels Bohr

CONTINUOUS RANGE

Continuous values

Classical physics allows energy to be released or absorbed in a continuous range of values. This is akin to a person walking up a ramp, in that the person can occupy any number of positions on the ramp.

Light can excite electrons to higher energy levels. They emit photons as they fall back to their ground state.

Emission spectra

Quantum physics can be seen in action in the emission spectra of atoms. When excited by photons of light, an atom's electrons gain energy, jumping up to a higher orbital (see p.79). As the electrons fall back to their relaxed ground state, they emit a photon. The amount of energy carried by the photon is quantized and always consistent, so the colour of emitted light is always the same.

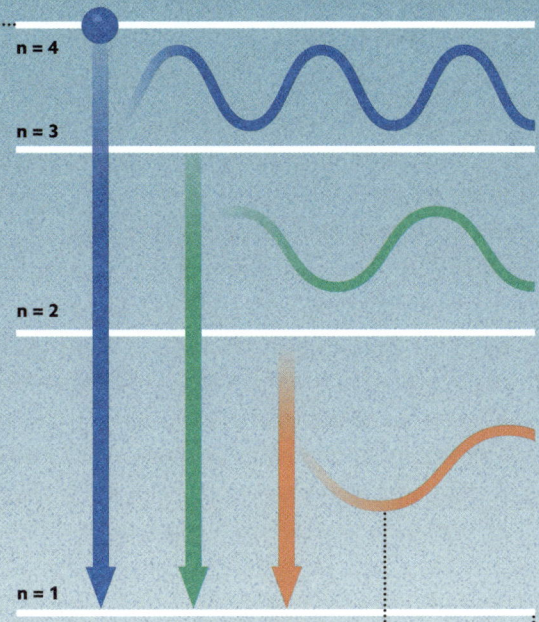

n = 4

n = 3

n = 2

n = 1

The wavelength (and therefore colour) of the emitted light can only have certain quantized values.

Energy level of electron orbital

Just as matter is quantized into "chunks", so are some types of energy.

ENERGY IN JUMPS

Quantized values

According to quantum theory, electromagnetic energy is quantized – it can only be released or absorbed in packets, or quanta, of a set value. This is like ascending a staircase with defined distances between the steps.

WAVES OF POSSIBILITY

At the quantum level, entities (such as electrons) that were previously thought of as discrete particles, localized in one place, also behave as waves; other entities (such as light) that were once thought of as spread-out waves also behave as particles. In quantum mechanics, a particle does not have a definite position and momentum; instead, its properties are spread out in a space defined by a wave function – a mathematical equation that gives a statistical prediction of where the particle will be at a given time. Once the particle is observed or interacts with something, the wave function changes or "collapses", giving one answer for the location of the particle.

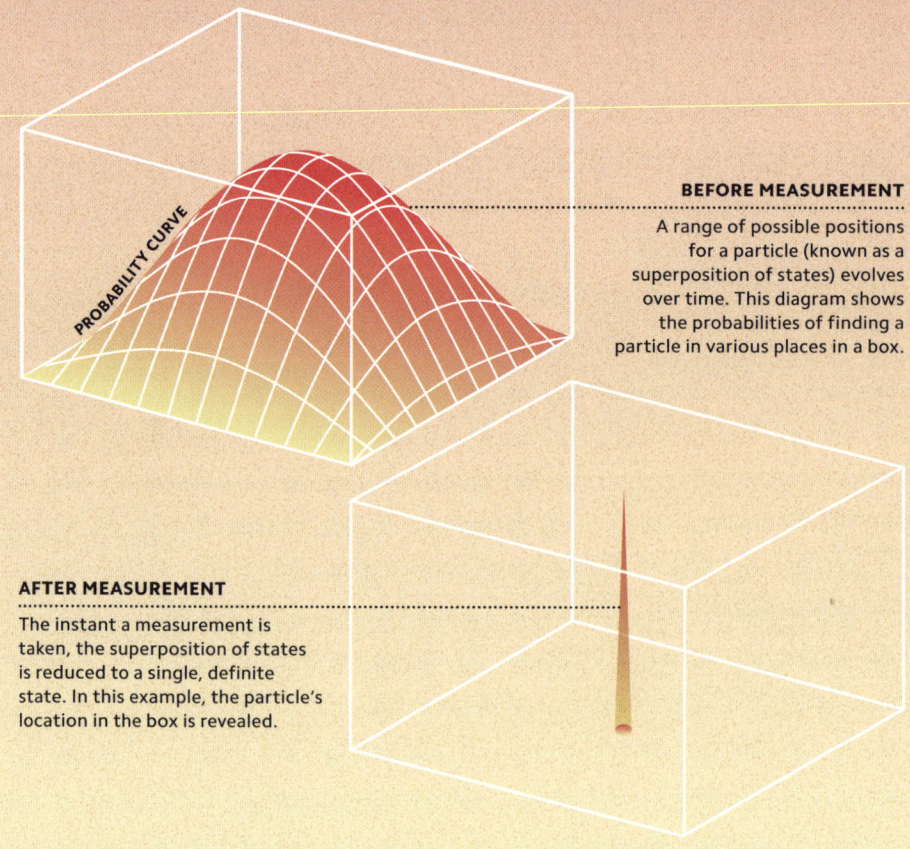

PROBABILITY CURVE

BEFORE MEASUREMENT
A range of possible positions for a particle (known as a superposition of states) evolves over time. This diagram shows the probabilities of finding a particle in various places in a box.

AFTER MEASUREMENT
The instant a measurement is taken, the superposition of states is reduced to a single, definite state. In this example, the particle's location in the box is revealed.

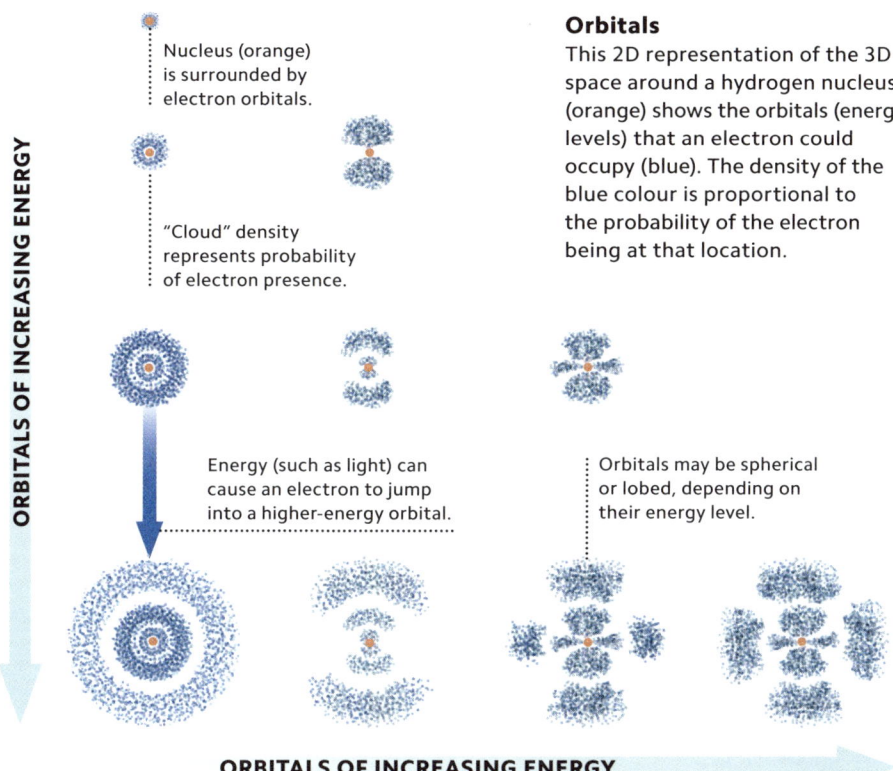

ORBITALS OF INCREASING ENERGY

Nucleus (orange) is surrounded by electron orbitals.

"Cloud" density represents probability of electron presence.

Energy (such as light) can cause an electron to jump into a higher-energy orbital.

Orbitals
This 2D representation of the 3D space around a hydrogen nucleus (orange) shows the orbitals (energy levels) that an electron could occupy (blue). The density of the blue colour is proportional to the probability of the electron being at that location.

Orbitals may be spherical or lobed, depending on their energy level.

ORBITALS OF INCREASING ENERGY

A NEW MODEL

Early models of atomic structure envisaged electrons circling the nucleus rather like planets moving around the Sun (see p.13). Quantum theory, however, explains the behaviour of electrons in a different way. Three-dimensional mathematical functions, called orbitals, describe the probability of an electron being at any particular point at a given time. Electrons in orbitals can only have certain (quantized) amounts of energy; the lowest-energy orbitals are occupied first. The highest-energy orbitals, furthest from the nucleus, contain the valence electrons that take part in chemical reactions (see p.19).

TIME, SPACE, AND MOTION

Put forward by Albert Einstein in 1905, the special theory of relativity explains the behaviour of objects that move close to the speed of light. It assumes two things: that the speed of light is constant, and that the laws of physics are the same in all inertial frames of reference (that is, where objects are stationary or move at constant velocity relative to one another). The idea is that intervals of time and space – long assumed to be "absolute" (that is, independent of observers) – are in fact relative quantities, and time and space are interwoven into a single continuum known as spacetime.

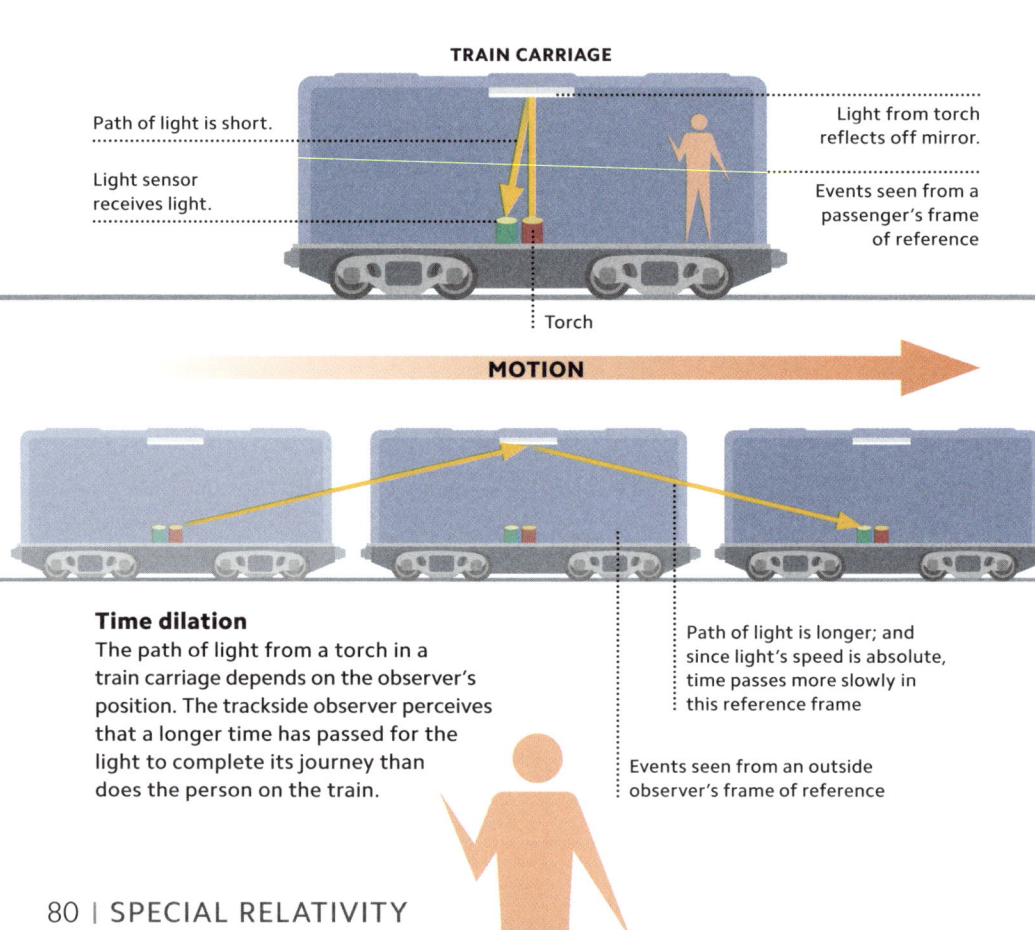

TRAIN CARRIAGE

Path of light is short.

Light sensor receives light.

Light from torch reflects off mirror.

Events seen from a passenger's frame of reference

Torch

MOTION

Time dilation
The path of light from a torch in a train carriage depends on the observer's position. The trackside observer perceives that a longer time has passed for the light to complete its journey than does the person on the train.

Path of light is longer; and since light's speed is absolute, time passes more slowly in this reference frame

Events seen from an outside observer's frame of reference

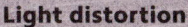

Light distortion
In this diagram, the grid of lines represents four-dimensional spacetime. Anything with mass will warp spacetime – the larger the mass, the greater the distortion. Gravity is the effect that this distortion has on other things; even light is affected.

Actual position of distant star

Apparent position of distant star as seen by observer

The gravity of a massive object bends the fabric of spacetime and causes less massive objects nearby to follow curved paths.

Light follows a curved path in distorted spacetime.

Imaginary lines (geodesics) represent the shortest distances between points in spacetime.

Observer's location

Massive star

WARPED SPACETIME

Developed by Einstein in 1915, the general theory of relativity explains the phenomenon of gravity more accurately than classical physics. Gravity is not considered to be a force; instead, the effects normally attributed to gravity arise because objects with mass or energy distort spacetime. By curving spacetime, two objects travelling in parallel straight lines move closer together – as if attracted by a force. The consequences of the theory are that intervals of time and space are different for different observers (as in special relativity) and time runs more slowly close to massive objects.

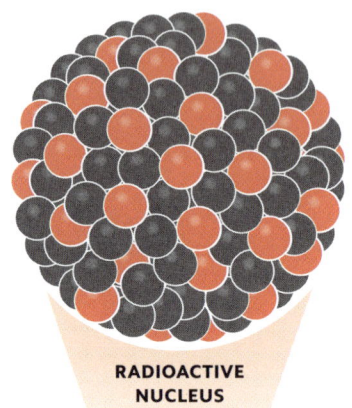

RADIOACTIVE NUCLEUS

Types of radiation
Unstable nuclei, such as uranium-235 (left), can emit different types of particles and radiation that have different energies. If these penetrate the human body, they can cause serious cell damage.

The half-life of uranium-235 is about 704 million years

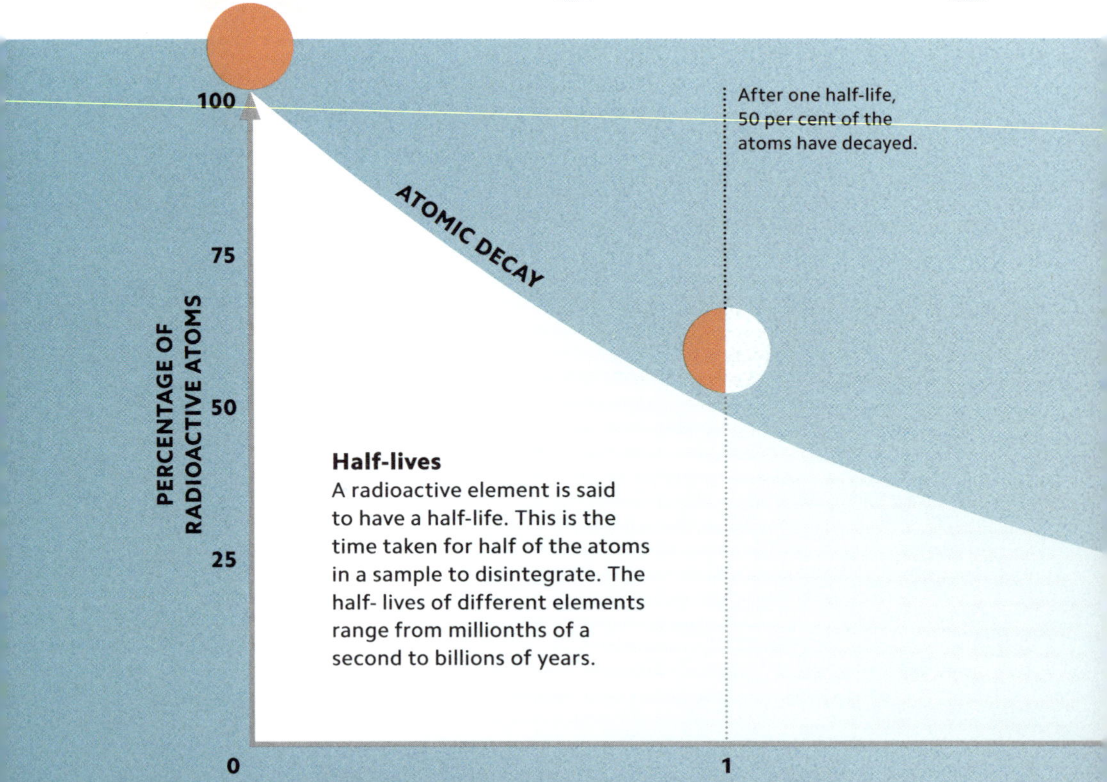

After one half-life, 50 per cent of the atoms have decayed.

ATOMIC DECAY

PERCENTAGE OF RADIOACTIVE ATOMS

100

75

50

25

0

1

Half-lives
A radioactive element is said to have a half-life. This is the time taken for half of the atoms in a sample to disintegrate. The half- lives of different elements range from millionths of a second to billions of years.

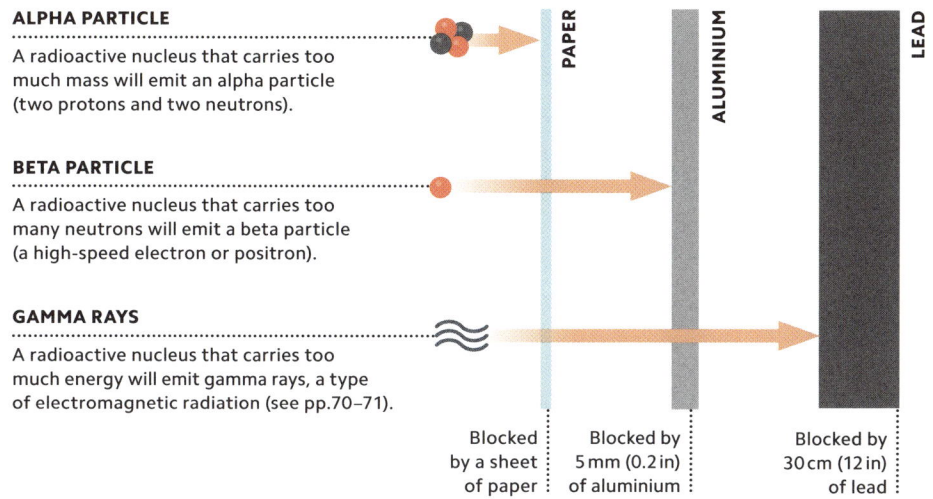

ALPHA PARTICLE

A radioactive nucleus that carries too much mass will emit an alpha particle (two protons and two neutrons).

BETA PARTICLE

A radioactive nucleus that carries too many neutrons will emit a beta particle (a high-speed electron or positron).

GAMMA RAYS

A radioactive nucleus that carries too much energy will emit gamma rays, a type of electromagnetic radiation (see pp.70–71).

PAPER

ALUMINIUM

LEAD

Blocked by a sheet of paper

Blocked by 5mm (0.2in) of aluminium

Blocked by 30cm (12in) of lead

NUCLEAR DECAY

The nucleus of an atom is said to be stable when the number of protons is in balance with the number of neutrons. Nuclei that are unstable will – over time – disintegrate, ejecting matter (particles) or energy to achieve a more stable configuration. This emission of energy or particles is called radioactivity. Some radioactive elements, such as uranium, thorium, and radon, occur naturally on Earth, while others, such as plutonium and americium, are made in nuclear reactors or explosions (see p.85).

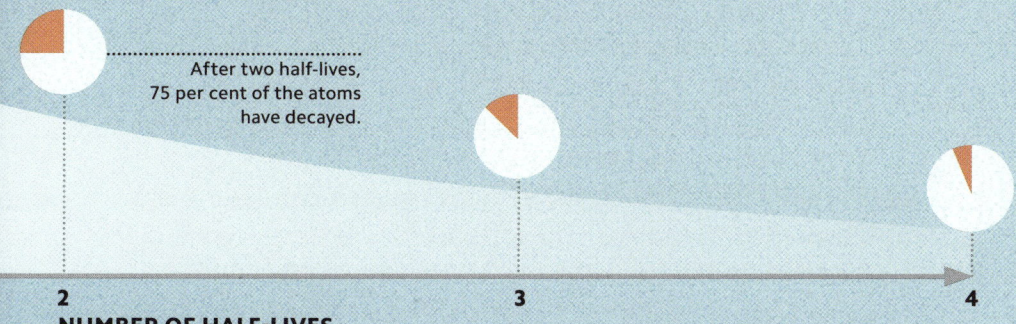

After two half-lives, 75 per cent of the atoms have decayed.

2   3   4

NUMBER OF HALF-LIVES

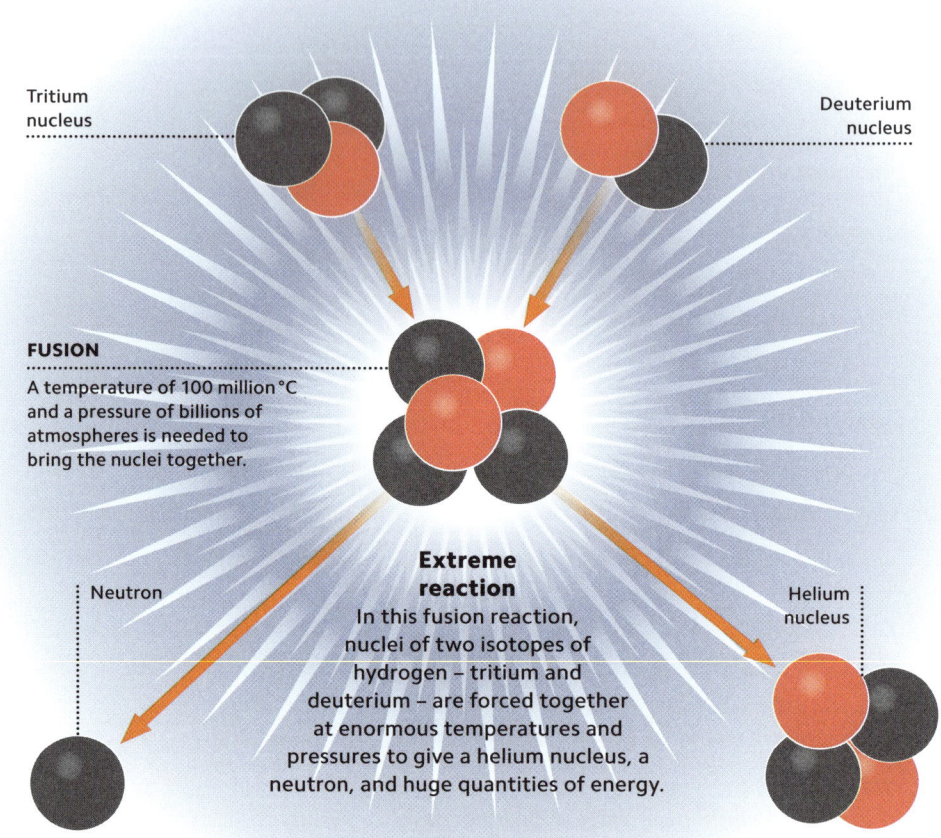

Tritium nucleus

Deuterium nucleus

FUSION
A temperature of 100 million °C and a pressure of billions of atmospheres is needed to bring the nuclei together.

Neutron

Helium nucleus

Extreme reaction
In this fusion reaction, nuclei of two isotopes of hydrogen – tritium and deuterium – are forced together at enormous temperatures and pressures to give a helium nucleus, a neutron, and huge quantities of energy.

FUSING ATOMS

Two light atomic nuclei, such as hydrogen nuclei, can – under extreme conditions – combine into a heavier nucleus, releasing huge amounts of energy. This process, called nuclear fusion, fuels stars, including our Sun, but has also been re-created by humans in thermonuclear bombs and in experimental fusion reactors. Fusion releases energy because the mass of the nuclei that come together in the reaction is greater than the mass of the products. The missing mass is converted into energy according to Einstein's famous equation, $E=mc^2$.

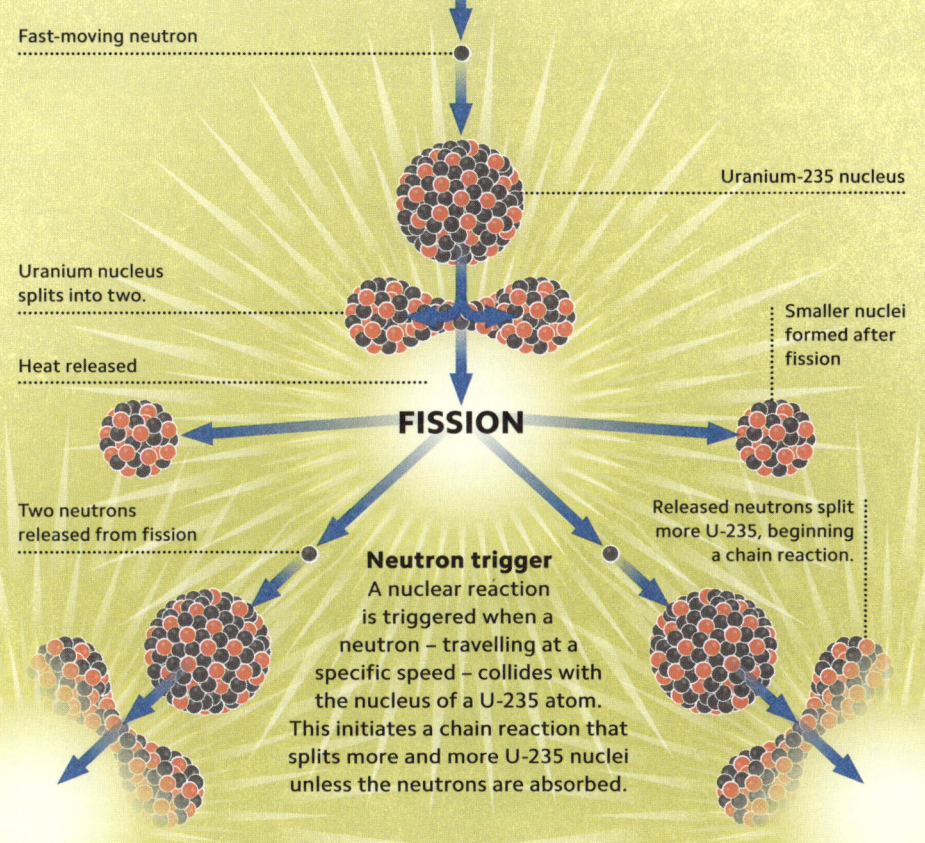

Fast-moving neutron

Uranium-235 nucleus

Uranium nucleus
splits into two.

Smaller nuclei
formed after
fission

Heat released

FISSION

Two neutrons
released from fission

Released neutrons split
more U-235, beginning
a chain reaction.

Neutron trigger
A nuclear reaction
is triggered when a
neutron – travelling at a
specific speed – collides with
the nucleus of a U-235 atom.
This initiates a chain reaction that
splits more and more U-235 nuclei
unless the neutrons are absorbed.

SPLITTING ATOMS

The nucleus of a heavy atom can become
unstable and split into two or more smaller
nuclei, releasing large amounts of heat energy.
This process, called nuclear fission, is the basis of
nuclear reactors and atomic weapons. It can, rarely,
occur naturally but is usually engineered by humans and
requires complex technology to harvest the resultant
energy. The fuel used for most nuclear reactions is an isotope
(see p.12) of uranium called U-235, which can be induced to
undergo fission relatively readily.

BIOLO

G Y

Life arose on our planet some 3.8 billion years ago and, driven by mutation and natural selection, it has evolved to occupy almost every available niche on Earth. Biology – the study of life – is accordingly an extremely broad field, encompassing the study of diversity, the relationships between organisms and their environment, and the structure and function of living things. Technology has enabled humans to probe organisms at the cellular and biochemical levels to the point that we can now examine the molecules that characterize life, and even intervene in heredity itself to design and produce novel organisms for the benefit of our own species.

LIVING THINGS

ORGANIZATION
Are made of basic building blocks called cells. Some have a single cell, others have billions.

RESPONSIVENESS
Change physically or chemically in response to outside stimuli.

GROWTH
Can increase in size or numbers.

ADAPTATION
A species can change over the longer term in response to a changing environment.

HOMEOSTASIS
Regulate the conditions within their bodies.

METABOLISM
Carry out chemical reactions that consume and release energy.

REPRODUCTION
Can make more of their own kind.

Philosophers and naturalists once thought that living things had a "vital force" that distinguished them from non-living things. We now know that no single property defines life but that all living things have a set of features (above) in common. On Earth, life is based on carbon (see p.21 and pp.100–101), but this may not hold true beyond our planet.

THE DIVERSITY OF LIFE

Life has existed on Earth for at least 3.8 billion years. The very first living things were prokaryotes – tiny organisms whose single cells were simple in structure. Estimates put the number of species on Earth today at more than 15 million, but so many microbes are yet to be described that the true number may be in the billions. The latest attempts to classify this diversity based on genetics and biochemistry place life into three domains – Bacteria, Archaea, and Eukaryotes – all of which are thought to have arisen from the same common ancestor.

COMMON ANCESTOR

PROKARYOTES
These single-celled organisms are around 0.1 to 5.0 micrometres across. They lack a nucleus and contain no membrane-bound substructures (organelles).

BACTERIA
Less than 2 micrometres across, bacteria are made up of a single cell that lacks a nucleus and membrane-bound organelles.

ARCHAEA
Similar to bacteria in size, these organisms have a very different biochemistry. Some gain energy by using sulfur, hydrogen, or ammonia.

EUKARYOTES
This domain contains the familiar kingdoms of organisms, such as plants and animals. Their cells are relatively large (5 to 100 micrometres across), and have a nucleus and membrane-bound organelles that carry out specific functions.

- **PROTISTS**
- **FUNGI**
- **PLANTS**
- **ANIMALS**

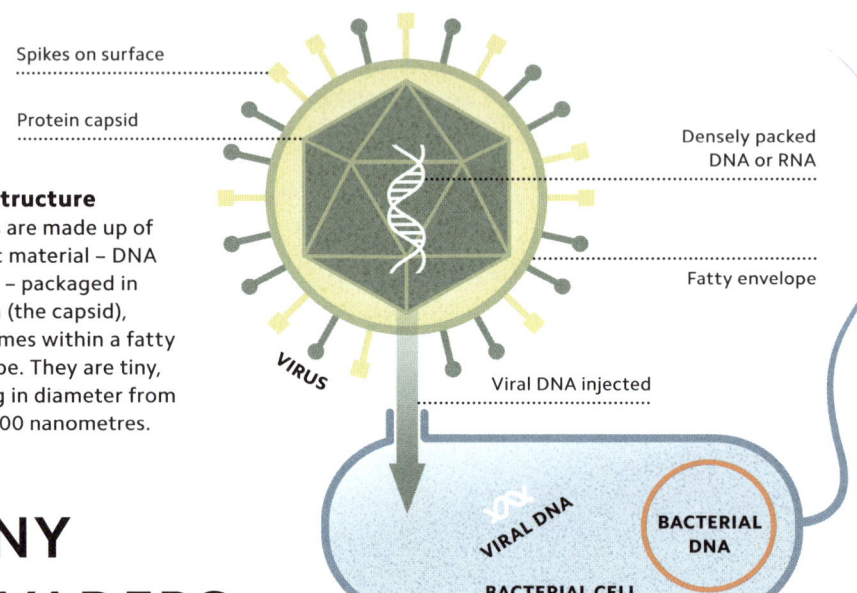

Spikes on surface

Protein capsid

Densely packed DNA or RNA

Fatty envelope

Viral structure

Viruses are made up of genetic material – DNA or RNA – packaged in protein (the capsid), sometimes within a fatty envelope. They are tiny, ranging in diameter from 20 to 800 nanometres.

VIRUS

Viral DNA injected

VIRAL DNA

BACTERIAL DNA

BACTERIAL CELL

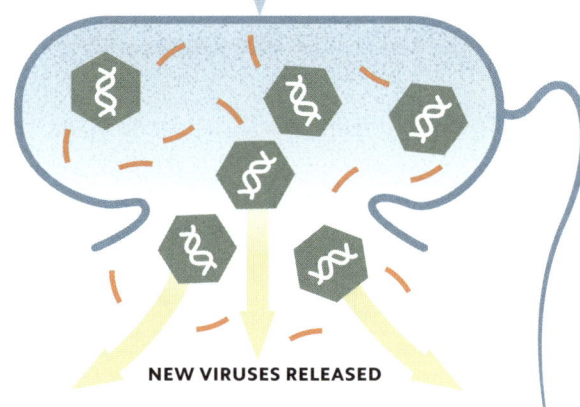

NEW VIRUSES RELEASED

TINY INVADERS

Viruses do not meet all the conditions that characterize life (see p.88). They have no cellular structure and cannot metabolize or reproduce on their own. To multiply they must invade a host – which may be an animal, plant, fungus, or bacterium – and hijack its cellular equipment. In doing so, viruses cause a range of animal diseases, including influenza and COVID-19, as well as plant diseases that can devastate important crops.

Viral infection

When viruses infect a cell (here, a bacterium) they attach to its surface and inject their DNA or RNA into the host. This genetic material directs the cell to make and assemble new viruses, which then burst out of the host.

UBIQUITOUS MICROBES

Bacteria are microscopic, single-celled prokaryotes (see p.89) that are best defined by what they lack. The cells contain no organelles – membrane-bound structures typical of the eukaryotic cells of plants and animals. Their genetic material is not arranged into chromosomes with a nucleus (as in animal and plant cells) but is instead held as a circular DNA molecule. Bacteria are almost ubiquitous on Earth. They are found in air, water, soil, and rocks, and within the bodies of other organisms; here, some cause disease, while others are essential to life.

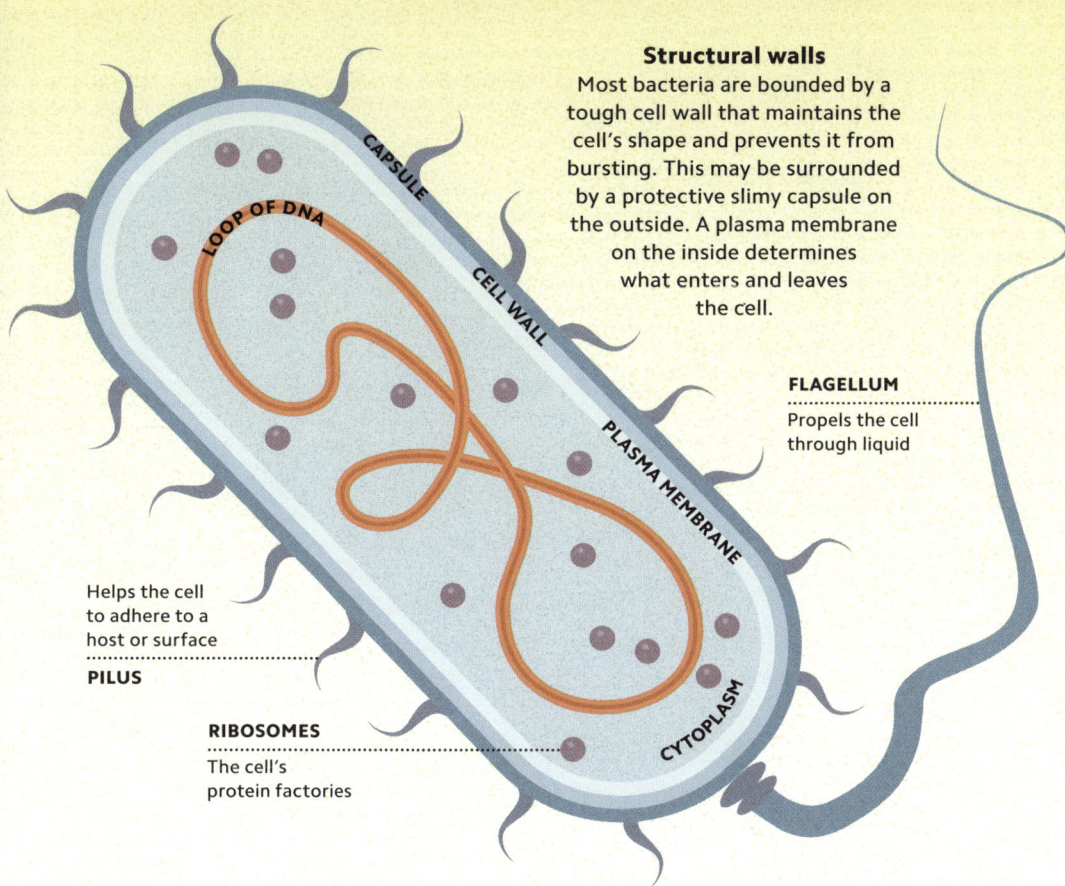

Structural walls
Most bacteria are bounded by a tough cell wall that maintains the cell's shape and prevents it from bursting. This may be surrounded by a protective slimy capsule on the outside. A plasma membrane on the inside determines what enters and leaves the cell.

CAPSULE

LOOP OF DNA

CELL WALL

PLASMA MEMBRANE

FLAGELLUM
Propels the cell through liquid

Helps the cell to adhere to a host or surface

PILUS

RIBOSOMES
The cell's protein factories

CYTOPLASM

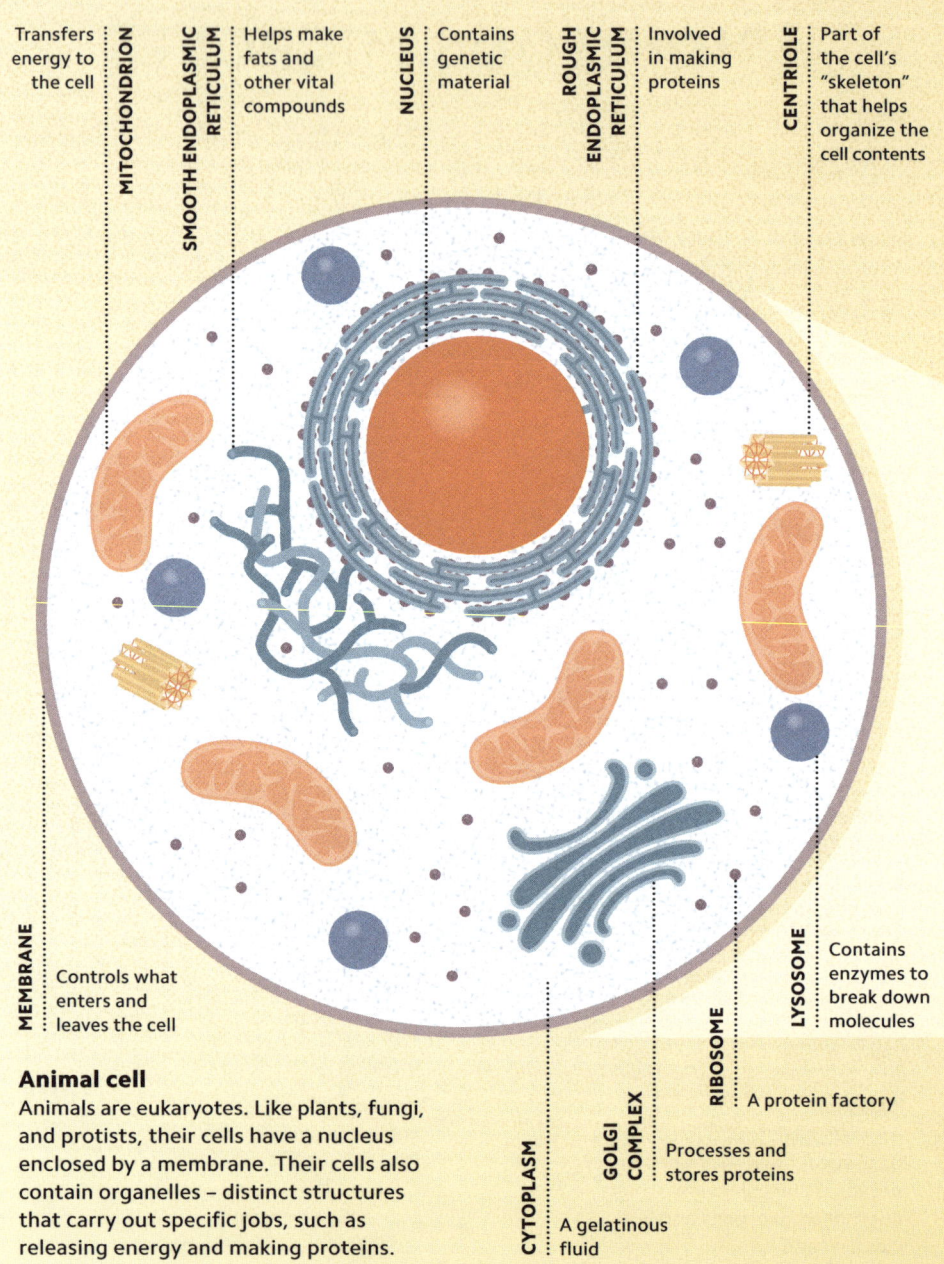

MITOCHONDRION — Transfers energy to the cell

SMOOTH ENDOPLASMIC RETICULUM — Helps make fats and other vital compounds

NUCLEUS — Contains genetic material

ROUGH ENDOPLASMIC RETICULUM — Involved in making proteins

CENTRIOLE — Part of the cell's "skeleton" that helps organize the cell contents

MEMBRANE — Controls what enters and leaves the cell

LYSOSOME — Contains enzymes to break down molecules

RIBOSOME — A protein factory

GOLGI COMPLEX — Processes and stores proteins

CYTOPLASM — A gelatinous fluid

Animal cell

Animals are eukaryotes. Like plants, fungi, and protists, their cells have a nucleus enclosed by a membrane. Their cells also contain organelles – distinct structures that carry out specific jobs, such as releasing energy and making proteins.

MAKING AN ANIMAL

There are around nine million animal species on our planet. Although they evolved from a single-celled ancestor, they are multicellular. Most are made up of millions or billions of eukaryotic cells. Animal cells of similar type come together to form tissues (see below) that perform specific roles. Two or more types of tissue are combined in organs, such as the heart, liver, and skin, which are the functional units of the animal body. Organs work together in systems – for example, the digestive system includes the mouth, oesophagus, liver, and intestines.

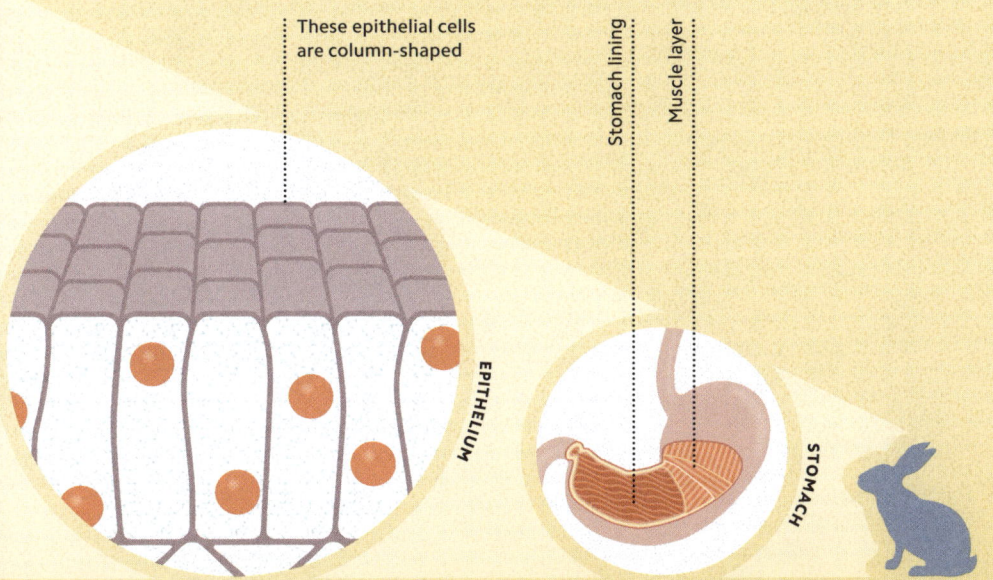

These epithelial cells are column-shaped

Stomach lining

Muscle layer

EPITHELIUM

STOMACH

Tissues
Animals have four basic types of tissue – muscle, connective, nervous, and epithelial (which lines internal organs and body cavities, and is found in the skin).

Organs
Various tissues come together to form organs. The stomach, for example, is lined with epithelial tissue, while surrounding muscle tissue churns the food within.

Plant cell

The cells of plants are supported by a rigid cell wall made mainly of cellulose. The cell's interior includes a large, membrane-bound storage organ called a vacuole. This is filled with a watery solution that exerts outward pressure, like the air in a balloon, so helping to keep the plant upright.

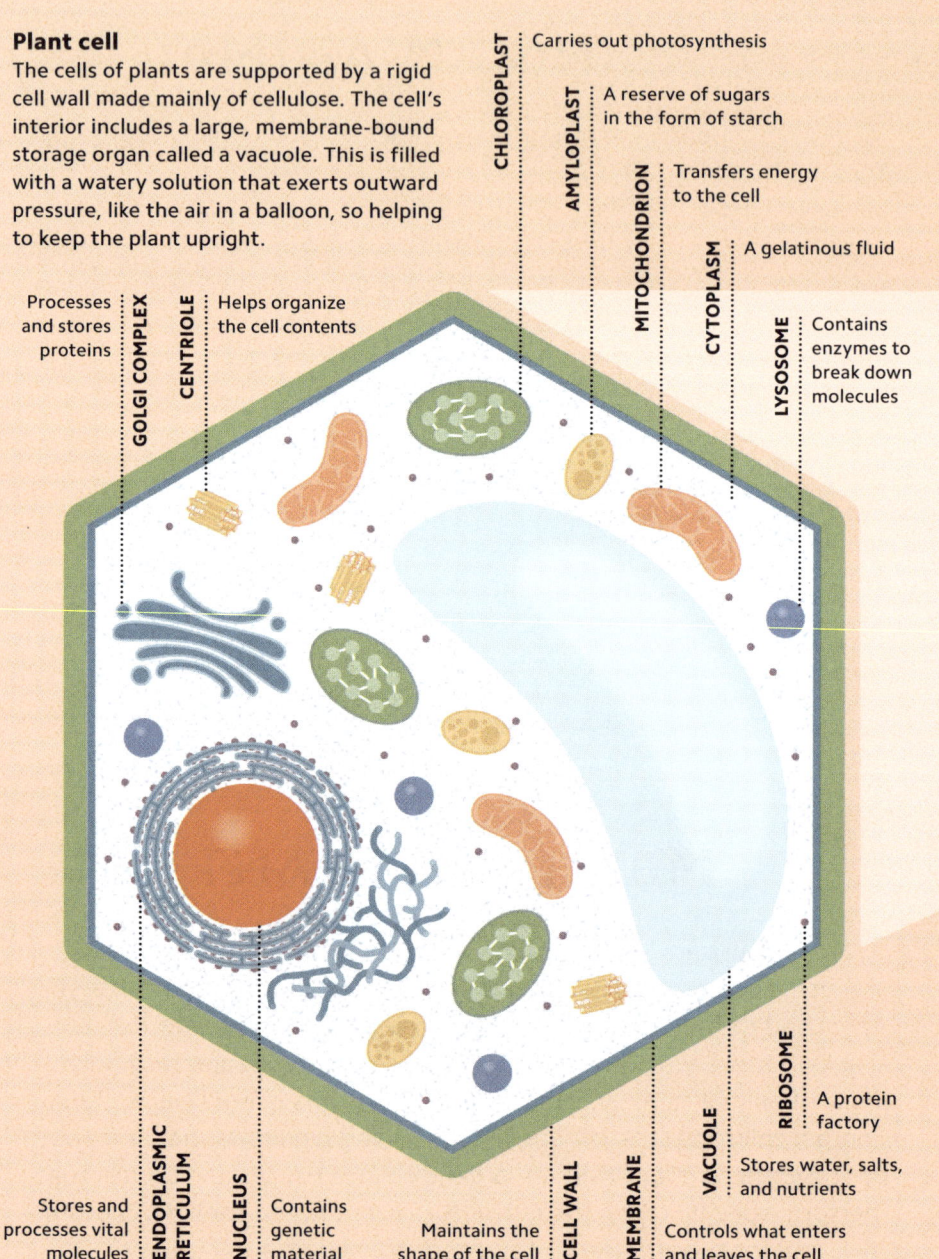

CHLOROPLAST Carries out photosynthesis

AMYLOPLAST A reserve of sugars in the form of starch

MITOCHONDRION Transfers energy to the cell

CYTOPLASM A gelatinous fluid

LYSOSOME Contains enzymes to break down molecules

GOLGI COMPLEX Processes and stores proteins

CENTRIOLE Helps organize the cell contents

ENDOPLASMIC RETICULUM Stores and processes vital molecules

NUCLEUS Contains genetic material

CELL WALL Maintains the shape of the cell

MEMBRANE Controls what enters and leaves the cell

VACUOLE Stores water, salts, and nutrients

RIBOSOME A protein factory

Tissues

Similar types of cells, organized into tissues, cooperate on a particular task in the body. Ground tissue comprises the bulk of a plant and is the site of most photosynthesis, while dermal tissue protects the surface of the plant.

Organs

A plant's organs – leaves, stems, and roots – consist of two or more tissue types. They can be modified for a particular function: for example, a potato is a modified stem devoted to the storage of food in the form of starch.

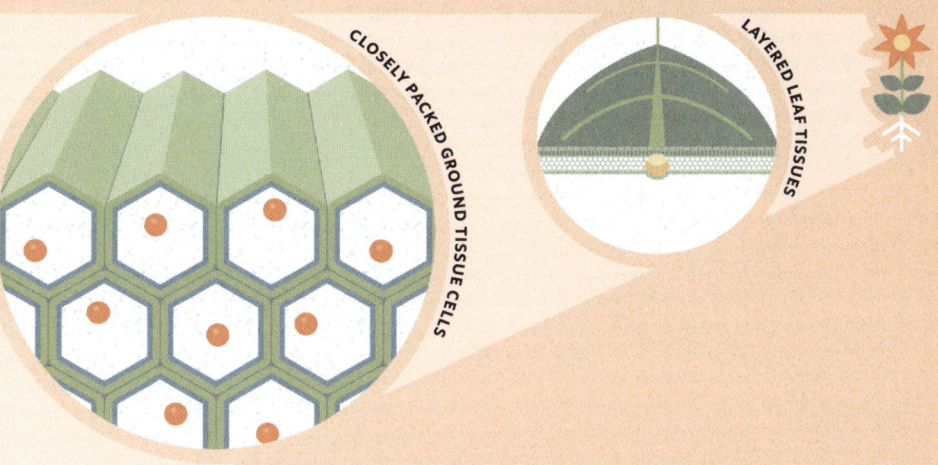

CLOSELY PACKED GROUND TISSUE CELLS

LAYERED LEAF TISSUES

MAKING A PLANT

Most plants are able to photosynthesize – to make sugars from water and carbon dioxide in the presence of light. These sugars are used to fuel growth and development. Plants are multicellular eukaryotes and their cells have many structures in common with those of animal cells. Unlike animal cells, they have rigid cell walls and contain chloroplasts – organelles containing pigments that carry out photosynthesis. Plant cells are organized into tissues and organs that perform specific tasks; their structure is optimized for light capture and for the storage and transport of sugars and nutrients.

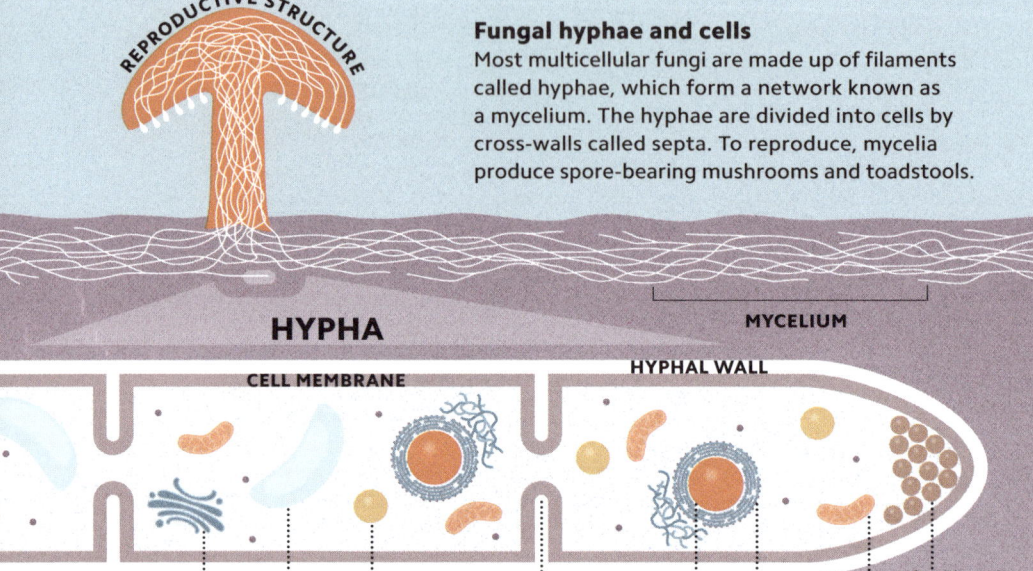

REPRODUCTIVE STRUCTURE

Fungal hyphae and cells
Most multicellular fungi are made up of filaments called hyphae, which form a network known as a mycelium. The hyphae are divided into cells by cross-walls called septa. To reproduce, mycelia produce spore-bearing mushrooms and toadstools.

HYPHA

MYCELIUM

CELL MEMBRANE

HYPHAL WALL

Processes and stores proteins · **GOLGI COMPLEX**

Breaks down and stores vital compounds · **VACUOLE**

LIPID BODIES · Store fats in the cell

Divides hyphae into cells · **SEPTUM**

Contains genetic material · **NUCLEUS**

Stores and processes molecules · **ENDOPLASMIC RETICULUM**

MITOCHONDRION · Transfers energy to the cell

VESICLES · Play a key role in the extension of hyphae

THREADS OF LIFE

Once grouped with plants, fungi are now classified in their own kingdom. Like plants, they are eukaryotes and have a cell wall, though this is made not of cellulose but chitin – a compound also found in the exoskeletons of insects. Unlike plants, they do not obtain food by photosynthesis: instead, they digest and absorb organic matter around them. Fungi make up a diverse group that includes multicellular forms, such as mushrooms and water moulds, and single-celled forms, such as yeasts.

Protozoa

These single-celled organisms feed by enveloping organic food particles.

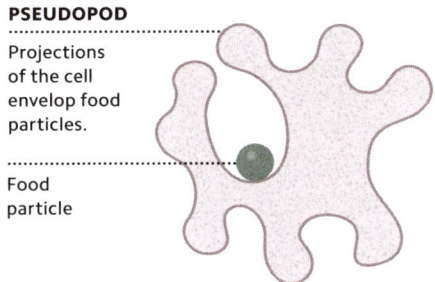

PSEUDOPOD
Projections of the cell envelop food particles.

Food particle

Algae

These aquatic, plantlike protists can be microscopic, such as diatoms (below), or very large, such as kelp.

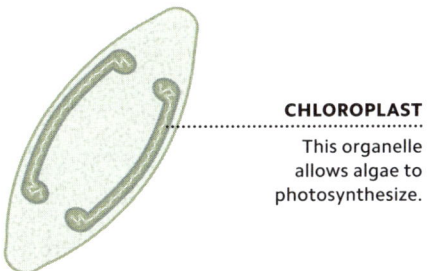

CHLOROPLAST
This organelle allows algae to photosynthesize.

SINGLE-CELLED LIFE

The fourth kingdom of eukaryotes – Protista – is a diverse collection of mostly single-celled organisms that cannot readily be classified as plants, animals, or fungi. Protists do not all share a common ancestor but evolved from various predecessors. The diversity of their forms is mirrored in their lifestyles: some photosynthesize, like plants; others envelop or digest organic material, like fungi and animals.

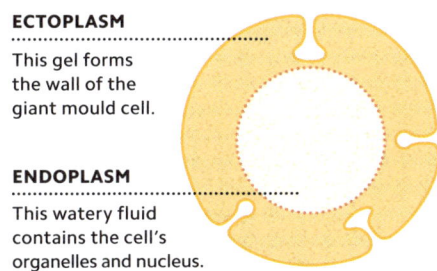

ECTOPLASM
This gel forms the wall of the giant mould cell.

ENDOPLASM
This watery fluid contains the cell's organelles and nucleus.

WATER PROPULSION
Two whiplike flagellae allow mould spores to swim.

Slime moulds

These tiny, fungus-like protists live in damp areas; some may unite to form "fruiting bodies" that produce spores.

Water moulds

Like fungi, these organisms form mycelia, which can be seen on decaying matter. Some, such as rusts, are plant pathogens.

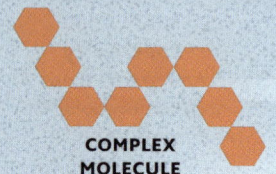

BREAKDOWN
CATABOLIC REACTION

COMPLEX MOLECULE

SIMPLE MOLECULES

Reactions that break molecules into simpler components, often releasing energy in the process, are called catabolic.

LIFE'S CHEMICAL PATHWAYS

ENERGY

The sum of all the chemical reactions that take place within the cells of a living thing is called metabolism. Many hundreds of individual reactions, often linked together in complex pathways, provide a cell with energy, build the molecules used for growth and repair, and break down parts of the cell for disposal. Metabolic pathways are remarkably similar in very different organisms, suggesting that they evolved long ago.

Cellular reactions that use energy to build complex molecules from simpler ones are called anabolic.

SIMPLE MOLECULES

ANABOLIC REACTION
SYNTHESIS

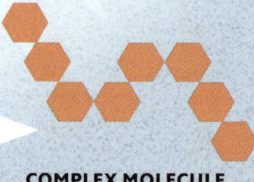

COMPLEX MOLECULE

SUBSTRATE

The molecule targeted by the action of the enzyme is called the substrate.

ENZYME

The substrate binds to one part of the enzyme, called the active site.

ACTIVE SITE

Specificity

An enzyme binds to a target molecule and forms an enzyme–substrate complex. The enzyme attaches at its active site, which is shaped to perfectly fit the target molecule.

ENZYME–SUBSTRATE COMPLEX

Binding with the enzyme forces the substrate into a shape that helps make – or, here, break – chemical bonds.

CATALYSTS OF LIFE

Enzymes are protein molecules made by living cells. Their purpose is to speed up chemical reactions in the cell that would otherwise take place so slowly that most life could not exist. Enzymes are catalysts (see p.30); they are not themselves consumed or altered by the reactions that they accelerate. They are highly specific, each one working only on one type of molecule in the cell; making more or less of a particular enzyme therefore directs the function of a particular cell. The number of enzymes produced by a single cell depends on the function of that cell. Overall, the human body produces thousands of enzymes.

PRODUCTS

The molecules that result from the reaction are released from the complex.

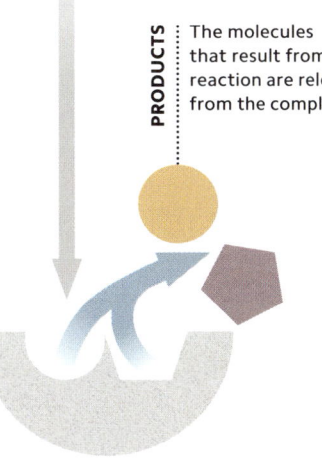

THE STUFF OF LIFE

Living organisms are made up of non-living carbon-based molecules, collectively called biomolecules. Some are small; simple sugars, for example, have a backbone of just six carbon atoms. However, many key biomolecules are large polymers containing many thousands or millions of carbon atoms. They fall into four major groups – carbohydrates, proteins, lipids, and nucleic acids.

SUBUNITS

GLUCOSE

AMINO ACIDS

Carbohydrates
These compounds of carbon, hydrogen, and oxygen include simple sugars as well as polymers such as cellulose, which makes up plant cell walls, and storage compounds including glycogen and starch. The starch molecule, for example, is made up of subunits of the sugar glucose.

Proteins
Proteins are polymers made up of small units called amino acids. Only 20 amino acids are found in living things. The number and sequence of amino acids in a protein determines its shape and function. Proteins help give cells structure and are involved in many metabolic processes.

POLYMER

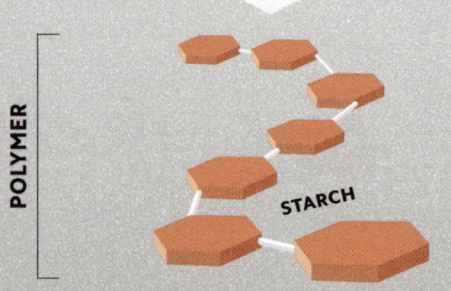

STARCH

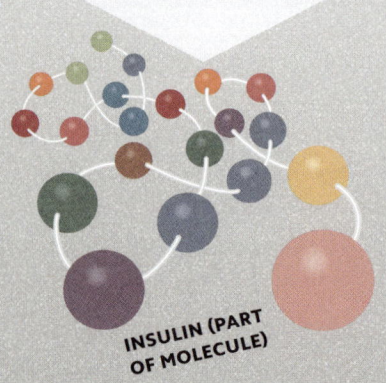

INSULIN (PART OF MOLECULE)

GLYCEROL

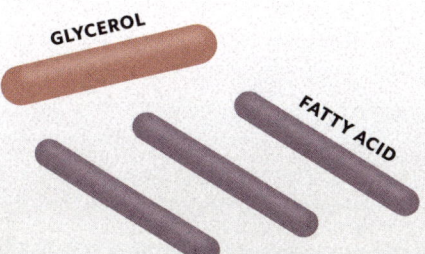

FATTY ACID

NUCLEOTIDES

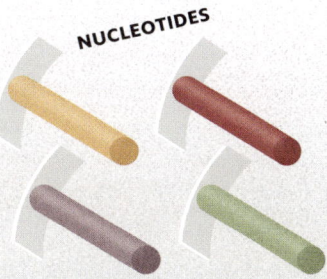

Lipids

This group includes fats, oils, phospholipids, waxes, and steroids. Their structures vary; fats are made up of a backbone of glycerol, to which are attached three long molecules of fatty acids. Lipids make up cell membranes and play a number of storage and metabolic roles.

Nucleic acids

These linear polymers store genetic information and direct the manufacture of proteins in the cell. There are two main types – deoxyribonucleic acid (DNA) and ribonucleic acid (RNA). The molecules are made up of a string of compounds called nucleotides.

TRIGLYCERIDE FAT

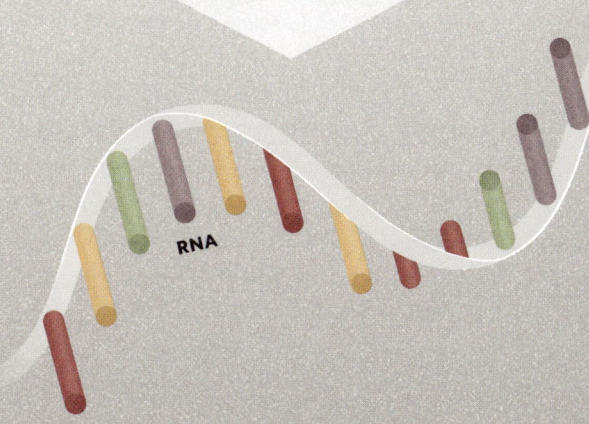

RNA

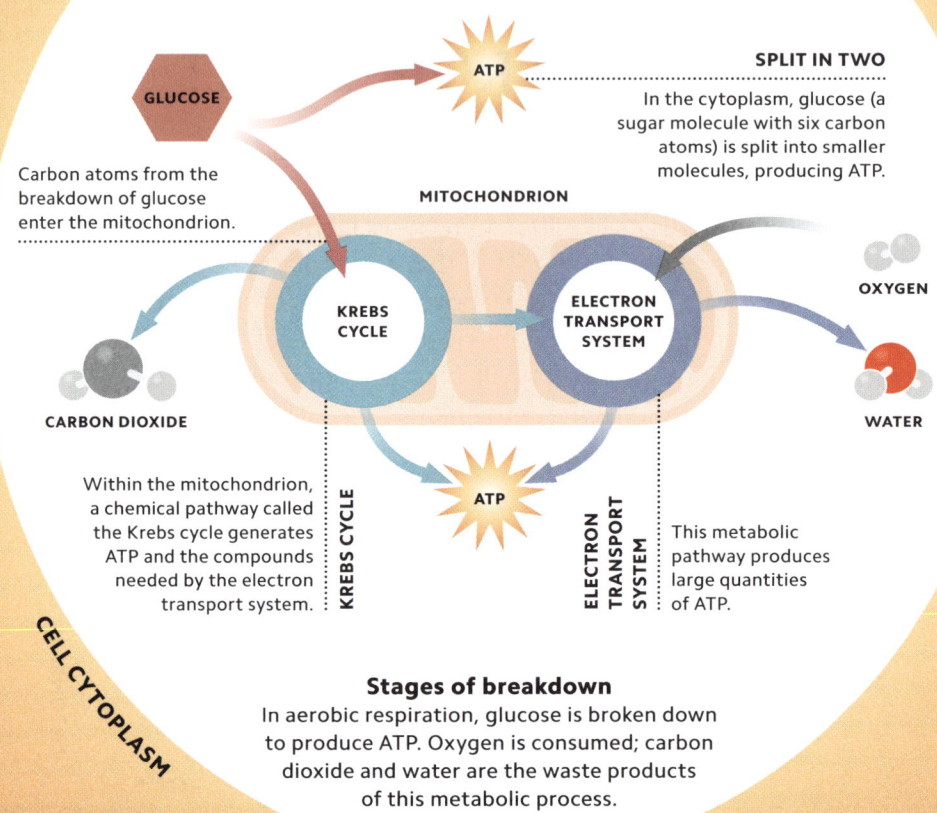

SPLIT IN TWO

In the cytoplasm, glucose (a sugar molecule with six carbon atoms) is split into smaller molecules, producing ATP.

ATP

GLUCOSE

Carbon atoms from the breakdown of glucose enter the mitochondrion.

MITOCHONDRION

KREBS CYCLE

ELECTRON TRANSPORT SYSTEM

OXYGEN

CARBON DIOXIDE

WATER

Within the mitochondrion, a chemical pathway called the Krebs cycle generates ATP and the compounds needed by the electron transport system.

KREBS CYCLE

ATP

ELECTRON TRANSPORT SYSTEM

This metabolic pathway produces large quantities of ATP.

CELL CYTOPLASM

Stages of breakdown

In aerobic respiration, glucose is broken down to produce ATP. Oxygen is consumed; carbon dioxide and water are the waste products of this metabolic process.

CELL POWER

In biology the word "respiration" describes the chain of enzyme-controlled chemical reactions that occur within all living cells to release energy from the sugar glucose. Some forms of respiration (aerobic) require oxygen; others (anaerobic) do not. Much of the process takes place in the cell's mitochondria; its products are carbon dioxide (which is removed and expelled from the cell), water, and chemical energy in the form of molecules of adenosine triphosphate (ATP), which is used to drive metabolic processes.

ENERGY FROM LIGHT

Plants, along with certain bacteria and algae, are able to harness light energy to produce their own food (glucose) from carbon dioxide and water. This process is called photosynthesis. Light energy is trapped by chlorophyll and other pigments within chloroplasts – structures in the cells of the organism. The metabolic pathways of photosynthesis are divided into two: light-dependent reactions that take place on folded membranes called thylakoids; and light-independent reactions that occur in the rest of the chloroplast.

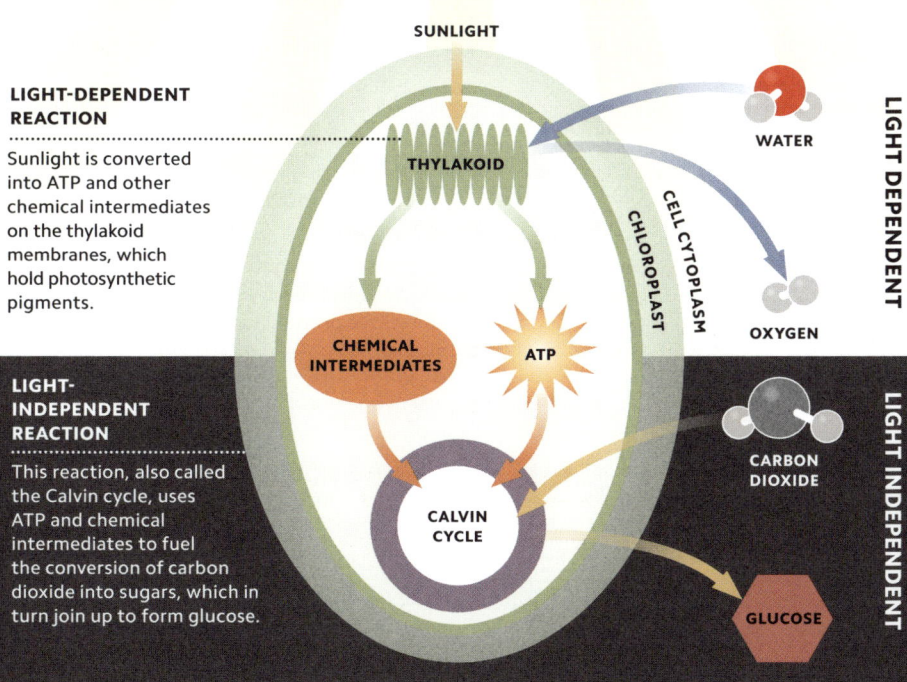

LIGHT-DEPENDENT REACTION

Sunlight is converted into ATP and other chemical intermediates on the thylakoid membranes, which hold photosynthetic pigments.

LIGHT-INDEPENDENT REACTION

This reaction, also called the Calvin cycle, uses ATP and chemical intermediates to fuel the conversion of carbon dioxide into sugars, which in turn join up to form glucose.

SUNLIGHT

THYLAKOID

CELL CYTOPLASM

CHLOROPLAST

WATER

OXYGEN

CHEMICAL INTERMEDIATES

ATP

CALVIN CYCLE

CARBON DIOXIDE

GLUCOSE

LIGHT DEPENDENT

LIGHT INDEPENDENT

Light and sugar

The leaves of green plants contain cells that have a dense concentration of chloroplasts – the organelles responsible for photosynthesis. Oxygen is a waste product of this metabolic process.

MOLECULES ON THE MOVE

Diffusion and osmosis are physical processes that distribute water and other vital molecules within and between cells. Molecules naturally move from areas of high to low concentration. This is diffusion, and it takes place in gases, liquids, and even solids. In osmosis, solvent molecules (usually water) move across a semi-permeable membrane, such as the cell membrane, to balance the concentration of a solute (such as glucose) on either side of the membrane.

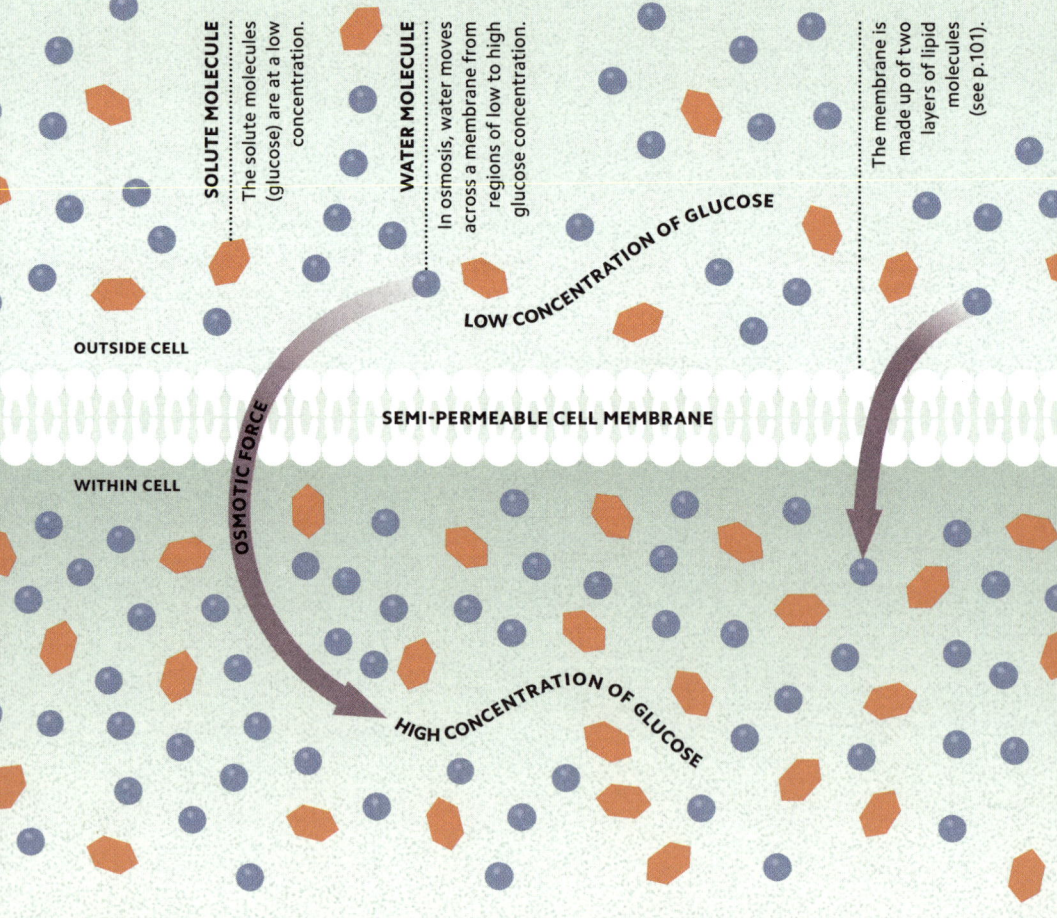

SOLUTE MOLECULE
The solute molecules (glucose) are at a low concentration.

WATER MOLECULE
In osmosis, water moves across a membrane from regions of low to high glucose concentration.

The membrane is made up of two layers of lipid molecules (see p.101).

LOW CONCENTRATION OF GLUCOSE

OUTSIDE CELL

OSMOTIC FORCE

SEMI-PERMEABLE CELL MEMBRANE

WITHIN CELL

HIGH CONCENTRATION OF GLUCOSE

CELL PUMPS

Cells take up some molecules through passive processes like diffusion and osmosis (see left). However, many vital molecules exist outside cells in lower concentrations than within, and so must be actively "pumped" in. A cell membrane is essentially a sheet of lipid molecules. Embedded in this sheet are large protein molecules that span the membrane. They are capable of binding to target molecules outside the cell and using energy, in the form of ATP (see p.102), to move them into the cell.

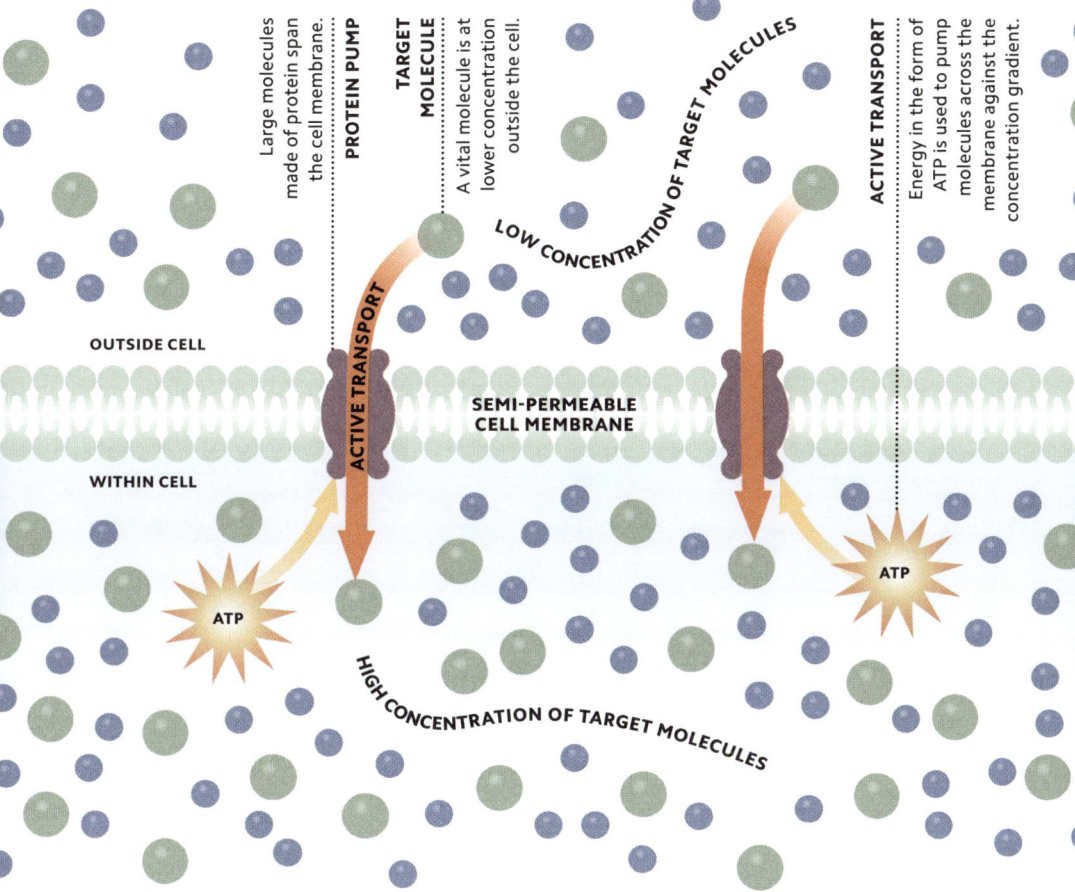

Large molecules made of protein span the cell membrane.

PROTEIN PUMP

TARGET MOLECULE

A vital molecule is at lower concentration outside the cell.

LOW CONCENTRATION OF TARGET MOLECULES

ACTIVE TRANSPORT

Energy in the form of ATP is used to pump molecules across the membrane against the concentration gradient.

OUTSIDE CELL

ACTIVE TRANSPORT

SEMI-PERMEABLE CELL MEMBRANE

WITHIN CELL

ATP

ATP

HIGH CONCENTRATION OF TARGET MOLECULES

Deoxyribonucleic acid (DNA) is a nucleic acid (see p.101) that carries genetic code – the information that directs a cell's function, growth, and reproduction. DNA is a chain of nucleotides, each one comprising a sugar (in this case deoxyribose), a phosphate group, and a nitrogen-rich base. The sugar and phosphate make up the "backbone" of the DNA molecule, while the bases, of which there are four – adenine (A), cytosine (C), guanine (G), and thymine (T) – project out from the backbone.

LONG DNA MOLECULE

LIFE LIBRARY

BASE PAIRS

Each base has a pairing partner; adenine always pairs up with thymine, and cytosine with guanine.

A particular sequence of bases along a DNA molecule makes up a gene; it encodes the instructions to produce a particular protein.

GENES ARE SEQUENCES

CYTOSINE GUANINE

ADENINE THYMINE

T A

G C

The double helix

DNA is made up of two long polymers twisted around one another into a double helix. Because of the consistent A–T and C–G pairing, one strand effectively mirrors the other. This helps maintain the integrity of the genetic code.

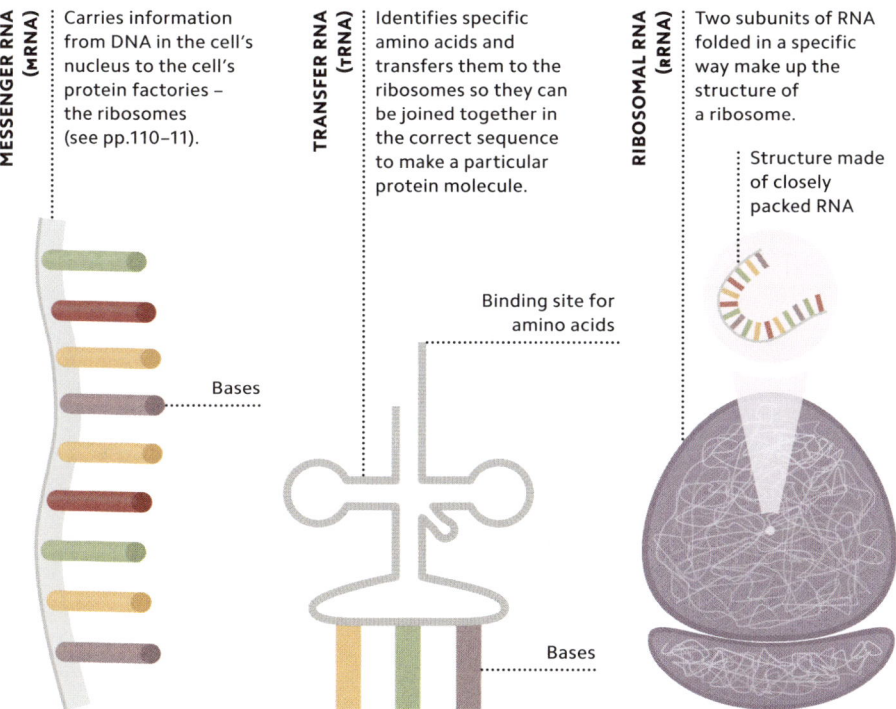

MESSENGER RNA (mRNA)
Carries information from DNA in the cell's nucleus to the cell's protein factories – the ribosomes (see pp.110–11).

TRANSFER RNA (tRNA)
Identifies specific amino acids and transfers them to the ribosomes so they can be joined together in the correct sequence to make a particular protein molecule.

RIBOSOMAL RNA (rRNA)
Two subunits of RNA folded in a specific way make up the structure of a ribosome.

Structure made of closely packed RNA

Binding site for amino acids

Bases

Bases

THE CELL'S EXECUTIVE

Ribonucleic acid is a nucleic acid (see p.101) – a chain of nucleotides composed of the sugar ribose, phosphate groups, and nitrogenous bases (see opposite). It is the primary carrier of information in viruses, but in most living organisms its main role is to execute the instructions held by a cell's DNA (see opposite). Like DNA it carries a four-letter code of bases, though the base uracil (U) replaces thymine (T); unlike DNA, it does not form a double helix, but can form strands or complex folded structures depending on its function. There are three main types of RNA – mRNA, tRNA, and rRNA – which perform different tasks in the cell.

PACKAGING DNA

In most eukaryotic cells, long DNA molecules are parcelled up into units called chromosomes within the nucleus. Humans have 46 chromosomes, but numbers vary between species. Organisms that reproduce sexually inherit chromosomes in pairs, with one half of each pair coming from each parent; humans, for example, have 22 pairs of "normal" chromosomes, plus two sex chromosomes (XX or XY) that determine biological sex.

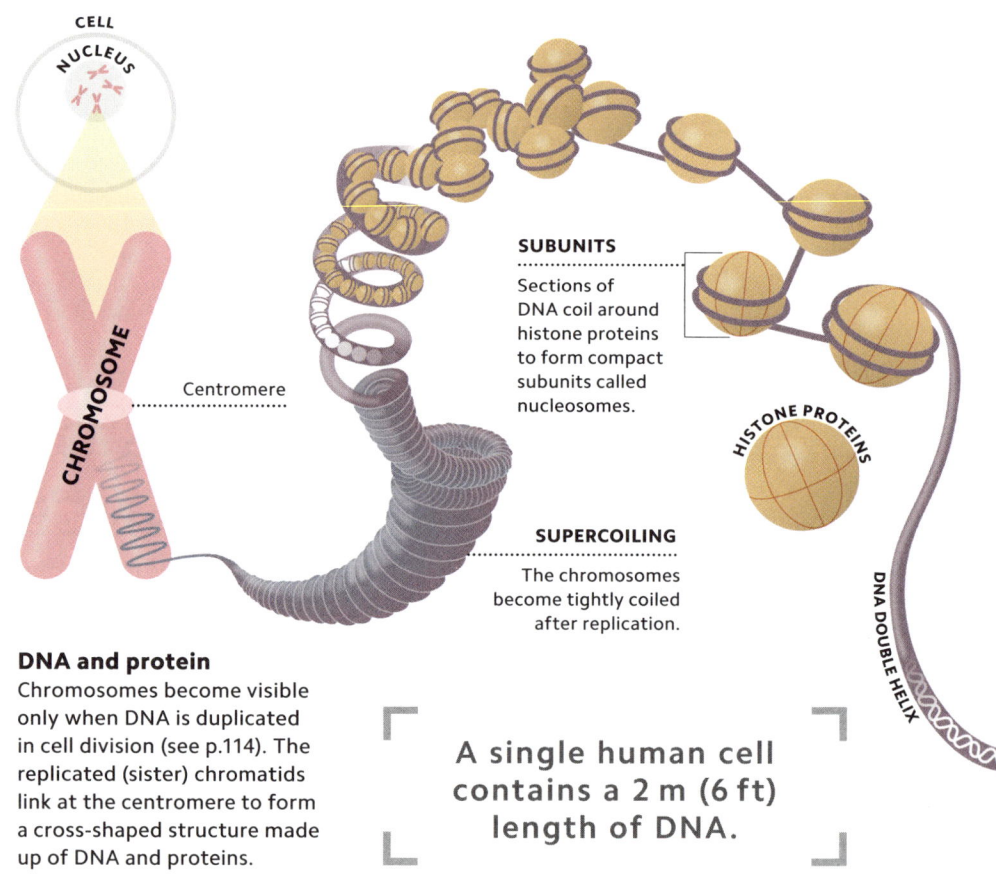

CELL

NUCLEUS

CHROMOSOME

Centromere

SUBUNITS

Sections of DNA coil around histone proteins to form compact subunits called nucleosomes.

HISTONE PROTEINS

SUPERCOILING

The chromosomes become tightly coiled after replication.

DNA DOUBLE HELIX

DNA and protein

Chromosomes become visible only when DNA is duplicated in cell division (see p.114). The replicated (sister) chromatids link at the centromere to form a cross-shaped structure made up of DNA and proteins.

A single human cell contains a 2 m (6 ft) length of DNA.

Structure of a gene

Promoter sections of a gene control whether it is switched on or off. Exons are sections that code for proteins. Introns are non-coding sections; these are removed during transcription (see p.110).

DNA is divided into coding and non-coding sections (exons and introns, respectively).

GENE

INTRON

EXON

INTRON

EXON

EXON

INTRON

EXON

INTRON

PROMOTER

DNA DOUBLE HELIX

UNITS OF INHERITANCE

A gene is a linear section of DNA that provides coded instructions for a cell to produce a particular protein. Organisms require thousands of different proteins to grow, survive, and reproduce, and so there are thousands of genes. Each protein is encoded by a particular sequence of the four "letters" of the genetic alphabet – the bases C, G, A, and T (see p.106) – along the DNA in the chromosomes. That sequence may be from several hundred to millions of bases long.

FROM GENE TO PROTEIN

Genes are sequences of DNA that determine which proteins a cell makes, and when. The process in which DNA directs the manufacture of proteins consists of two phases. In the first phase, transcription, the DNA is "read" and converted into mRNA (see p.107). In the next step – translation – this messenger molecule leaves the nucleus and becomes attached to ribosomes, where it is translated into a sequence of amino acids – the constituents of a protein.

Transcription

When a gene is switched on, the DNA section corresponding to that gene "unzips", allowing special enzymes to copy the gene into a strand of mRNA.

The strand of mRNA leaves the cell nucleus through pores in the nuclear membrane.

The DNA double helix "unzips" in the region of the activated gene, exposing both strands of the DNA molecule.

Enzymes work to copy the sequence of bases on the DNA strand into an RNA molecule.

DNA STRAND

MRNA

RNA NUCLEOTIDE

DNA DOUBLE HELIX

The mRNA strand grows as more nucleotides are added to it.

Nucleotides corresponding to the four genetic "letters" are linked in correct sequence.

NUCLEUS

FREED tRNA

Once used, tRNA molecules are free to collect more amino acid.

The protein assembled is a sequence of amino acids.

PROTEIN BUILDING

PROTEIN

AMINO ACID

TRANSPORT

Individual amino acids are brought to the ribosome by tRNA molecules.

TRNA

BASE RECOGNITION

The bases on the tRNA recognized the corresponding bases on the mRNA strand.

NUCLEAR MEMBRANE

MRNA

RIBOSOME

CYTOPLASM

READING mRNA

A mRNA strand is "read" sequentially as a ribosome moves down its length.

MOVING FACTORIES

The ribosomes are "protein factories" that move along a strand of mRNA.

Translation

The mRNA molecules become attached to ribosomes. Another form of RNA called tRNA now comes into play (see p.107). At one end of a tRNA molecule is a sequence of three bases; this corresponds to a three-letter "instruction" on the mRNA molecule. At the other end is a specific amino acid. As the ribosome moves along the mRNA molecule, enzymes link together the amino acids brought into the correct sequence by the tRNA molecules.

Agouti gene suppression

Agouti mice are fat and yellow, yet they occasionally produce offspring that are thin and brown. The number of brown offspring increases when pregnant mice are given a vitamin B12 supplement. This is because B12 contains chemical subunits called methyl groups that attach to certain bases on the DNA, making them "unreadable".

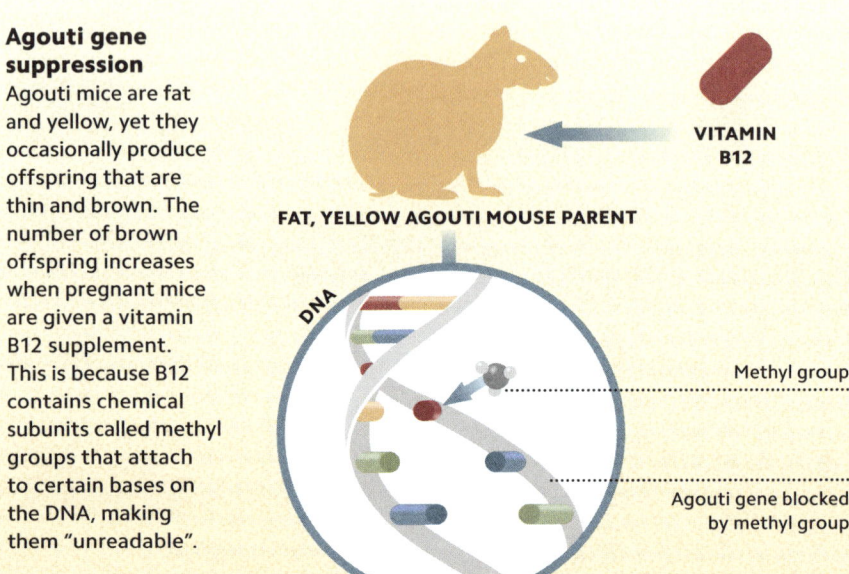

VITAMIN B12

FAT, YELLOW AGOUTI MOUSE PARENT

DNA

Methyl group

Agouti gene blocked by methyl group

THIN, BROWN AGOUTI MOUSE OFFSPRING

NATURE PLUS NURTURE

The sequence of DNA in an organism's cells is not the only thing that determines its characteristics. External factors, such as diet, behaviour, and environment, can all affect genetic material, changing which genes are switched on and off, and the ways in which they are expressed. These epigenetic factors (*epi* meaning "above" in Greek) can act on DNA, inducing chemical changes that affect transcription (see p.110) without altering the sequence of bases on the DNA strand. These epigenetic changes can be passed on from one generation to the next, but can sometimes also be reversed in response to environmental change.

PUTTING BIOLOGY TO WORK

Humans use a wide range of techniques to manipulate biological systems into making useful products. The term "biotechnology" describes these processes, which are as diverse as brewing (which harnesses yeast to make alcohol), DNA cloning, manufacturing biofuels, and plant breeding. In recent years, the field of genetic engineering has opened up countless new possibilities for changing the genomes of bacteria, plants, and animals.

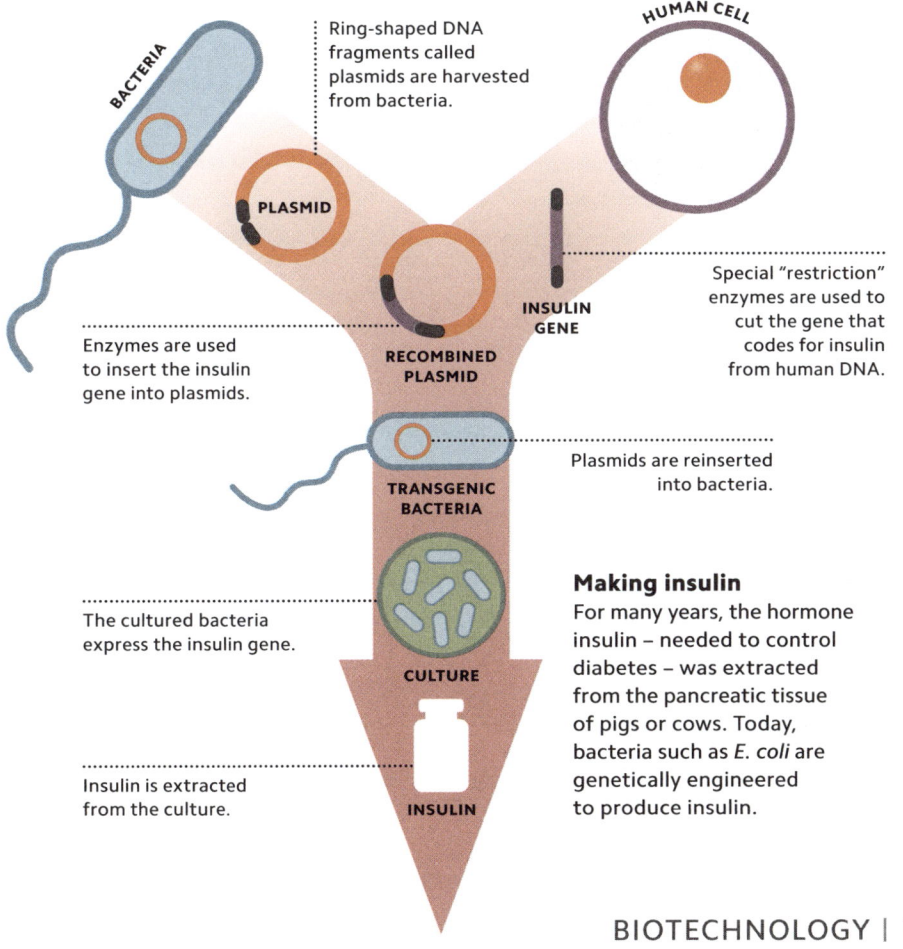

BACTERIA

Ring-shaped DNA fragments called plasmids are harvested from bacteria.

HUMAN CELL

PLASMID

INSULIN GENE

Special "restriction" enzymes are used to cut the gene that codes for insulin from human DNA.

RECOMBINED PLASMID

Enzymes are used to insert the insulin gene into plasmids.

TRANSGENIC BACTERIA

Plasmids are reinserted into bacteria.

The cultured bacteria express the insulin gene.

CULTURE

Insulin is extracted from the culture.

INSULIN

Making insulin

For many years, the hormone insulin – needed to control diabetes – was extracted from the pancreatic tissue of pigs or cows. Today, bacteria such as *E. coli* are genetically engineered to produce insulin.

DUPLICATING CELLS

When organisms grow and repair their bodies, their cells divide to increase in number. The cells resulting from such division must have the same genetic composition as the parent cell, with no additions, deletions, or changes. The process of ensuring the accurate transmission of the parent cell's DNA to its two daughter cells is called mitosis and takes place in almost every living cell. The parent cell's chromosomes are duplicated and allocated to the daughter cells in a precise series of steps. The cell's other contents, such as cytoplasm, mitochondria, and other organelles, are subsequently split between the daughter cells.

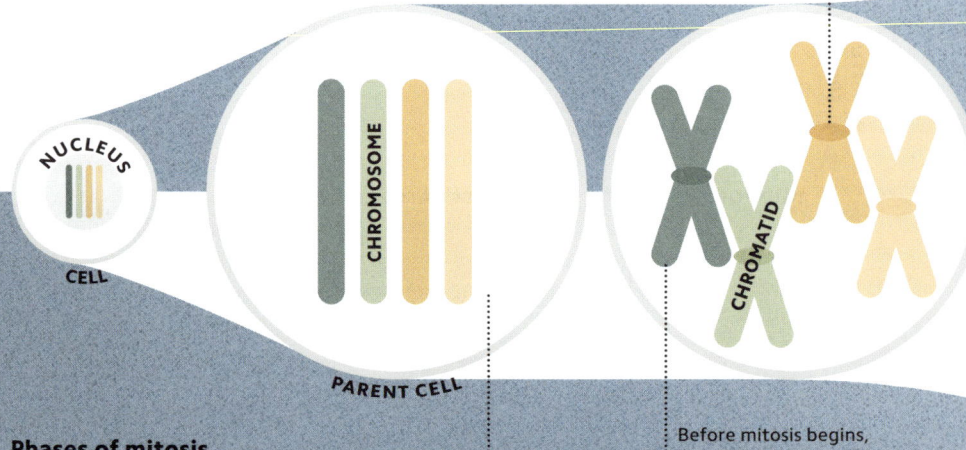

NUCLEUS

CELL

CHROMOSOME

PARENT CELL

Centromere

CHROMATID

Phases of mitosis

Here, a cell with just four chromosomes shows the process of mitosis. Most cells have many more chromosomes (humans have 46), making the coordination of cellular activity more complex.

MEMBRANE DISSOLVES

The nuclear membrane disappears.

DUPLICATION

Before mitosis begins, the cell's chromosomes are duplicated; each copy is now known as a chromatid. The two chromatids are joined together by a thickened section called the centromere. It is at this stage that chromosomes become visible in the cell.

> "Mitosis adds cells, while programmed cell death removes them."
> H. Robert Horvitz

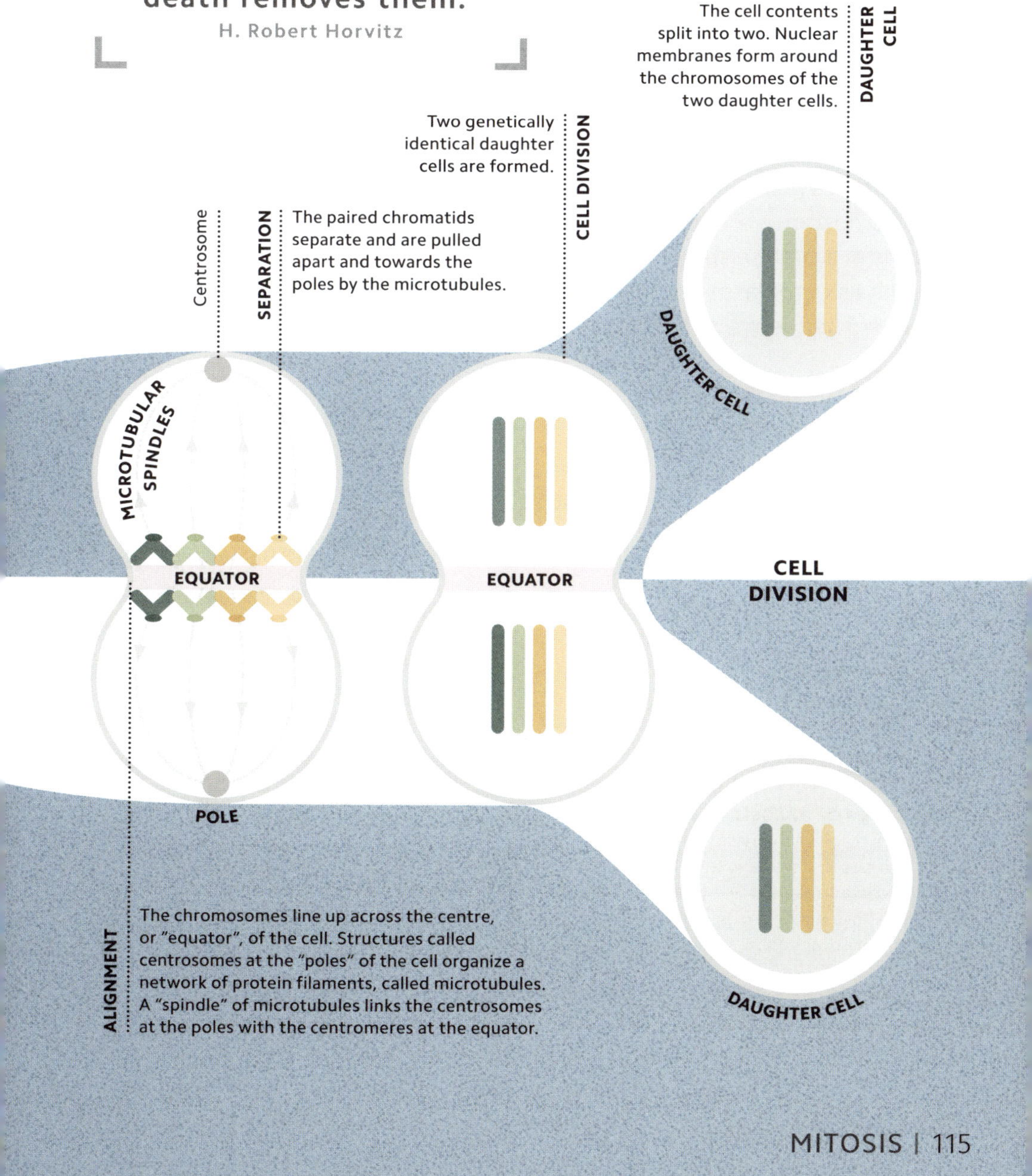

DAUGHTER CELL

The cell contents split into two. Nuclear membranes form around the chromosomes of the two daughter cells.

CELL DIVISION

Two genetically identical daughter cells are formed.

SEPARATION

The paired chromatids separate and are pulled apart and towards the poles by the microtubules.

Centrosome

DAUGHTER CELL

MICROTUBULAR SPINDLES

EQUATOR

EQUATOR

CELL DIVISION

POLE

ALIGNMENT

The chromosomes line up across the centre, or "equator", of the cell. Structures called centrosomes at the "poles" of the cell organize a network of protein filaments, called microtubules. A "spindle" of microtubules links the centrosomes at the poles with the centromeres at the equator.

DAUGHTER CELL

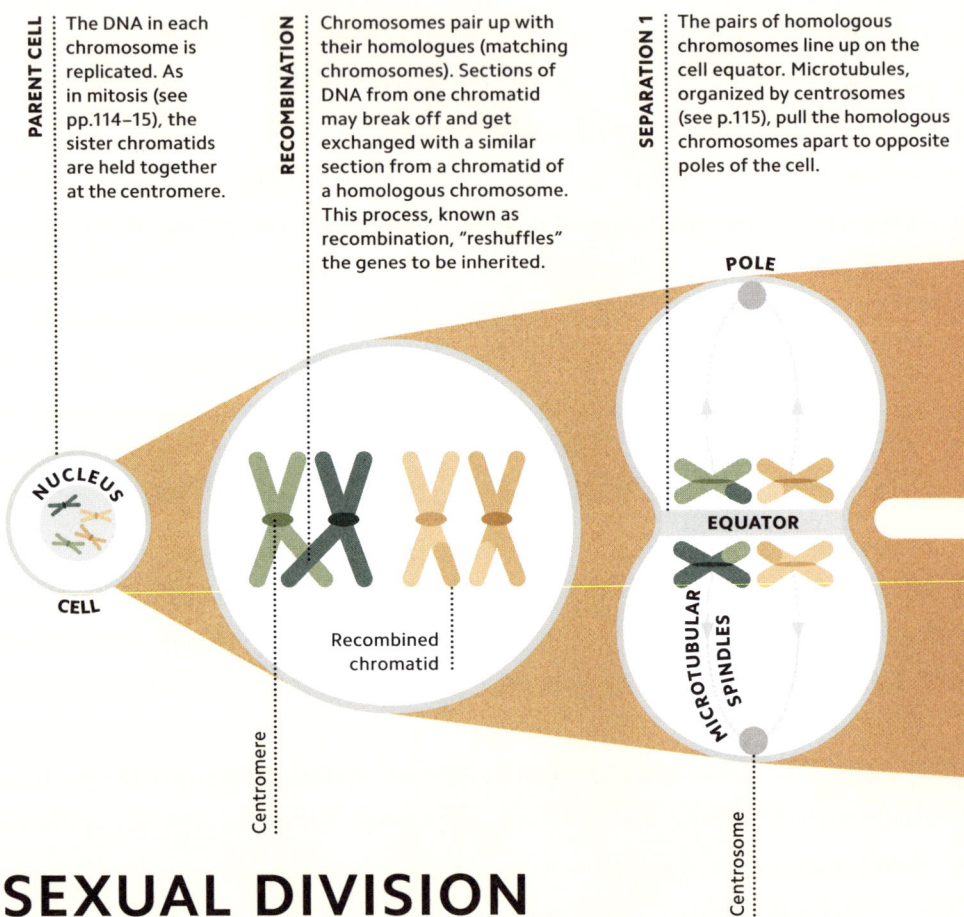

The DNA in each chromosome is replicated. As in mitosis (see pp.114–15), the sister chromatids are held together at the centromere.

Chromosomes pair up with their homologues (matching chromosomes). Sections of DNA from one chromatid may break off and get exchanged with a similar section from a chromatid of a homologous chromosome. This process, known as recombination, "reshuffles" the genes to be inherited.

The pairs of homologous chromosomes line up on the cell equator. Microtubules, organized by centrosomes (see p.115), pull the homologous chromosomes apart to opposite poles of the cell.

POLE

NUCLEUS

CELL

EQUATOR

Recombined chromatid

Centromere

MICROTUBULAR SPINDLES

Centrosome

SEXUAL DIVISION

Meiosis is a type of cell division that occurs only in sexual reproduction. A parent cell divides to produce four daughter cells, each of which has half the number of chromosomes of the parent. These cells form the gametes (the sperm and eggs in animals, and the pollen and ovules in plants). Gametes from two individuals later fuse during fertilization to produce a zygote with a full set of chromosomes. Crucially, meiosis allows genes on the chromosomes to be "reshuffled" and so creates variation – key to the process of natural selection (see p.121).

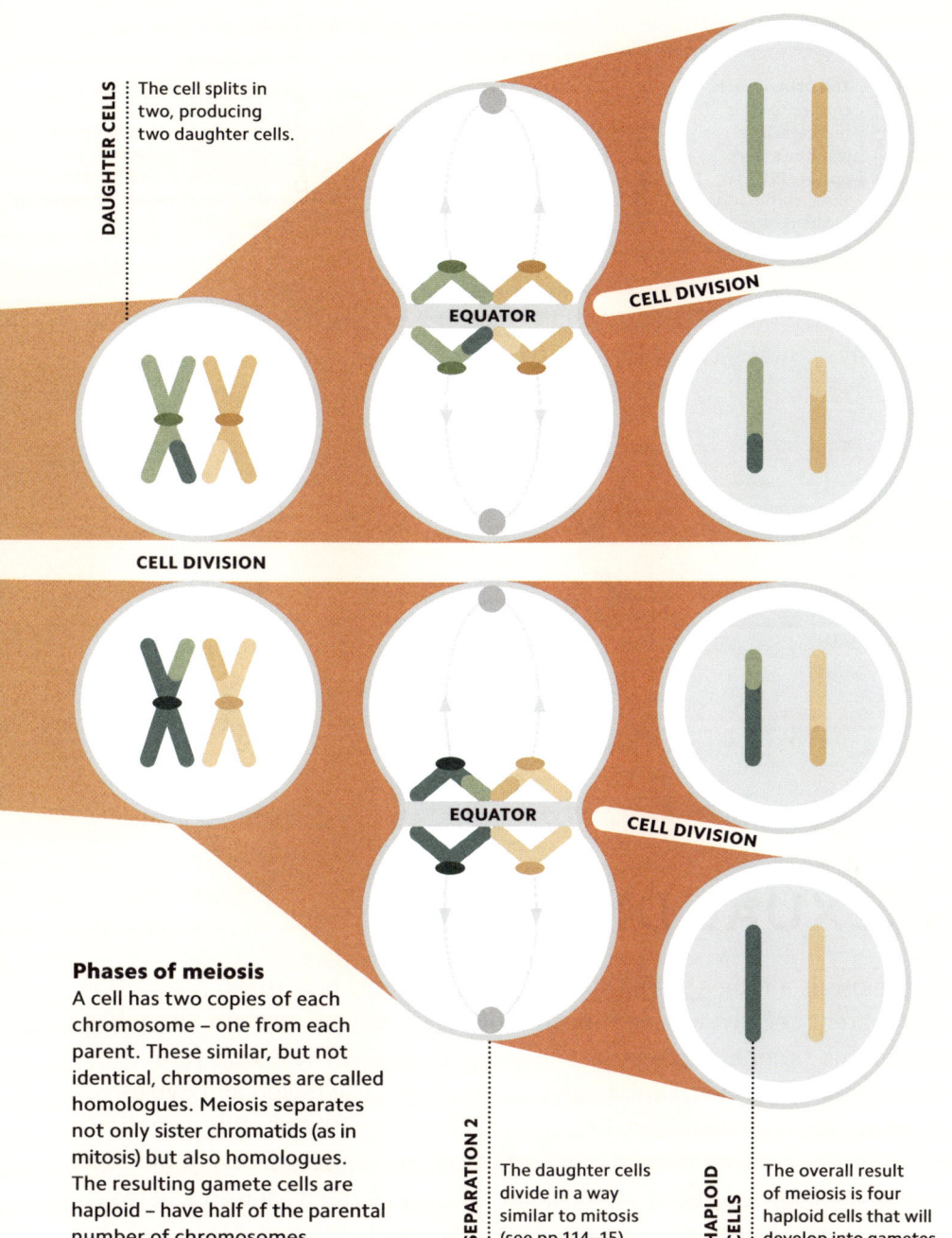

The cell splits in two, producing two daughter cells.

CELL DIVISION

EQUATOR

CELL DIVISION

EQUATOR

CELL DIVISION

Phases of meiosis
A cell has two copies of each chromosome – one from each parent. These similar, but not identical, chromosomes are called homologues. Meiosis separates not only sister chromatids (as in mitosis) but also homologues. The resulting gamete cells are haploid – have half of the parental number of chromosomes.

SEPARATION 2
The daughter cells divide in a way similar to mitosis (see pp.114–15).

HAPLOID CELLS
The overall result of meiosis is four haploid cells that will develop into gametes.

HOW GENES COMBINE

Most organisms have two sets of chromosomes. One set is inherited from each parent in sexual reproduction. This means that cells carry two copies of each gene. There are multiple "flavours" of each gene; these different forms are called alleles. The combination of alleles in an individual defines its genetic makeup, or genotype, which in turn determines its characteristics, or phenotype. Some alleles mask the effects of other alleles and are said to be dominant.

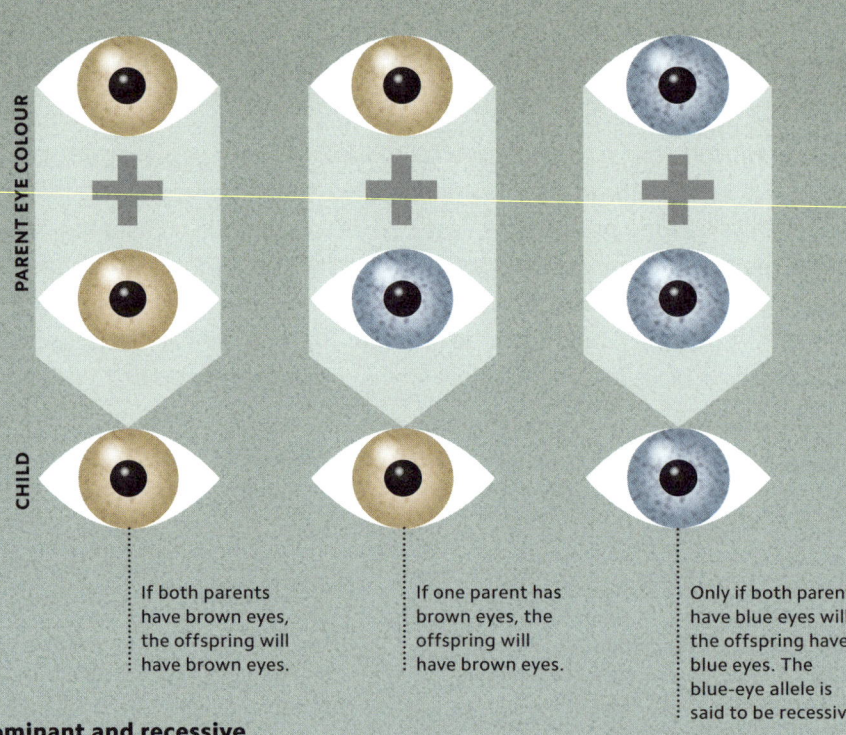

PARENT EYE COLOUR

CHILD

If both parents have brown eyes, the offspring will have brown eyes.

If one parent has brown eyes, the offspring will have brown eyes.

Only if both parents have blue eyes will the offspring have blue eyes. The blue-eye allele is said to be recessive.

Dominant and recessive
Many different genes and alleles determine pigmentation in human beings. One gene for eye colour, for example, exists as two alleles. One allele produces brown eyes, one blue eyes. However, the allele for brown eyes is dominant over the allele for blue eyes.

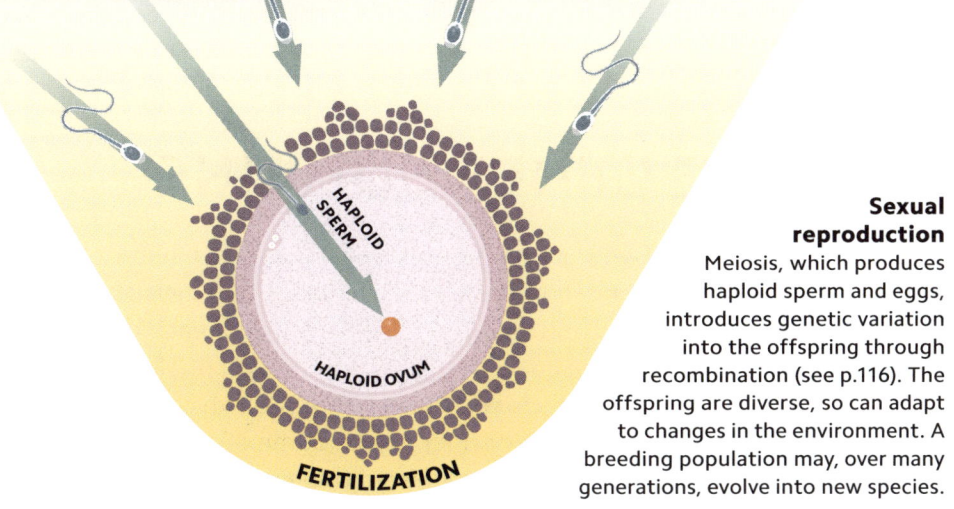

Sexual reproduction
Meiosis, which produces haploid sperm and eggs, introduces genetic variation into the offspring through recombination (see p.116). The offspring are diverse, so can adapt to changes in the environment. A breeding population may, over many generations, evolve into new species.

MEANS OF MULTIPLICATION

Sexual reproduction requires the formation of gametes (such as sperm and eggs in animals, and pollen and ovules in plants) by meiosis – a type of cell division (see pp.116–17). Fertilization occurs when male and female gametes come together, and a new organism develops with genes inherited from both parents. Asexual reproduction does not require meiosis. A new organism may be formed, for example, when a parent splits in two or part of it buds off into a distinct individual.

> "The flower is the poetry of reproduction."
> Jean Giraudoux

Asexual reproduction
The offspring of asexual reproduction are genetically identical to their parents. They can usually be produced rapidly, meaning that they can quickly fill suitable habitats, but they are far less able to adapt to changing conditions.

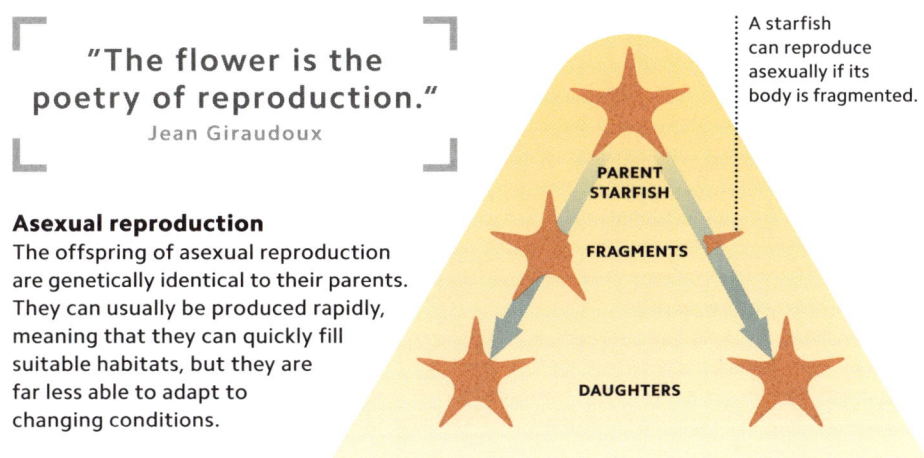

A starfish can reproduce asexually if its body is fragmented.

PARENT STARFISH

FRAGMENTS

DAUGHTERS

GENES, INDIVIDUALS, AND POPULATIONS

The entire set of genes possessed by an organism is known as its genome. Many genes exist in different variants, or alleles (see p.118). The particular set of alleles in an individual is what makes it genetically distinct from others in its species and is called its genotype. There are many different alleles of many genes, so the total number of permutations of alleles within a species is vast. The collection of genes and alleles available in all the individuals within one breeding population is known as the gene pool.

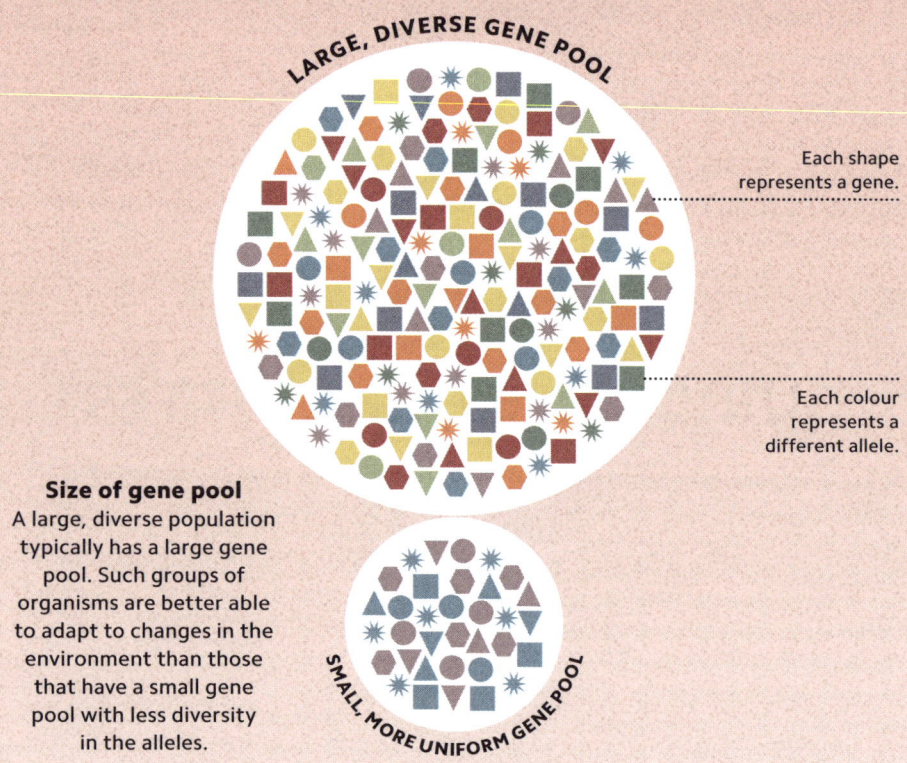

LARGE, DIVERSE GENE POOL

Each shape represents a gene.

Each colour represents a different allele.

SMALL, MORE UNIFORM GENE POOL

Size of gene pool
A large, diverse population typically has a large gene pool. Such groups of organisms are better able to adapt to changes in the environment than those that have a small gene pool with less diversity in the alleles.

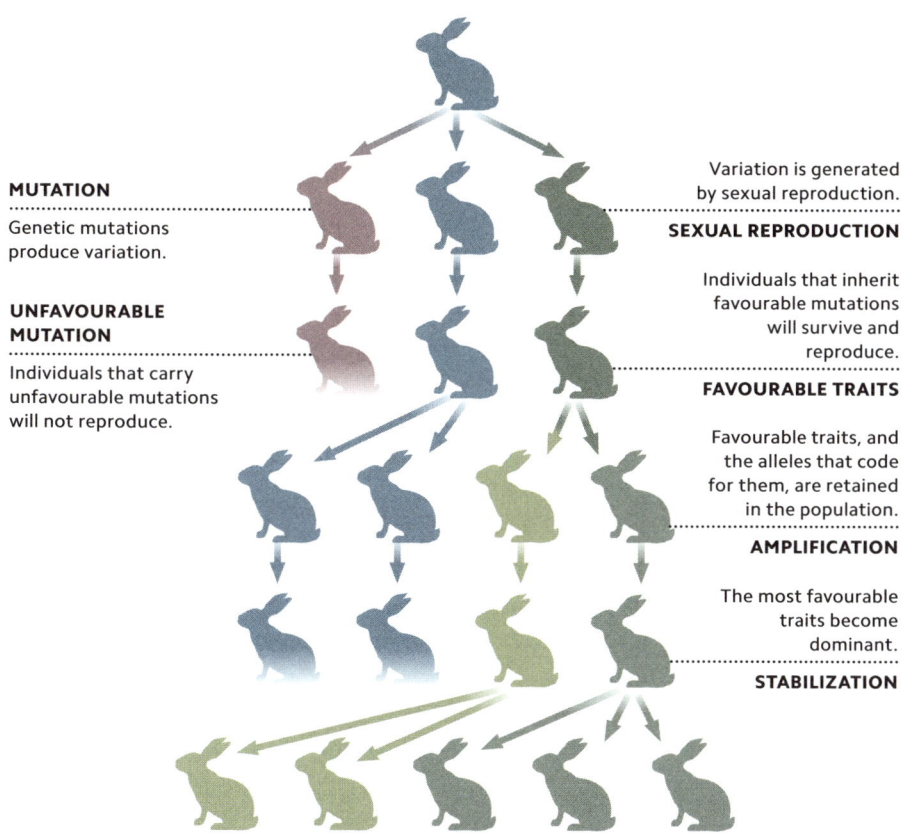

MUTATION

Genetic mutations produce variation.

UNFAVOURABLE MUTATION

Individuals that carry unfavourable mutations will not reproduce.

Variation is generated by sexual reproduction.

SEXUAL REPRODUCTION

Individuals that inherit favourable mutations will survive and reproduce.

FAVOURABLE TRAITS

Favourable traits, and the alleles that code for them, are retained in the population.

AMPLIFICATION

The most favourable traits become dominant.

STABILIZATION

SURVIVAL OF THE FITTEST

Living organisms adapt to changing environments through natural selection, the engine that drives evolution and the formation of new species (see p.122). Organisms within a population vary, and more variation is continually created by genetic mutations – mistakes in the DNA made when cells divide – and sexual reproduction, which "reshuffles" alleles (see p.116). Some individuals have traits that make them more likely to survive in their environment and so reproduce. These traits and their alleles are more likely to be passed on to the next generation.

BIRTH OF A SPECIES

A species is a group of organisms that can successfully interbreed. New species may form when a population becomes split into isolated groups. This happens if, for example, part of the population migrates away or is cut off by a natural barrier. Over time, the two groups may evolve by natural selection (see p.121) along different paths until they are no longer able to interbreed to produce fertile offspring. At this point, speciation has occurred. New species can also form in the same location as the parent species; this is more common in plants than in animals.

PARENT SPECIES
....
Individuals (represented here by dots) within a particular species are able to breed with one another.

BARRIER FORMATION
....
A natural barrier, such as a developing river, begins to split the population in two.

ISOLATED POPULATIONS
....
Natural selection drives the evolution of different characteristics in the two populations, which eventually become separate species.

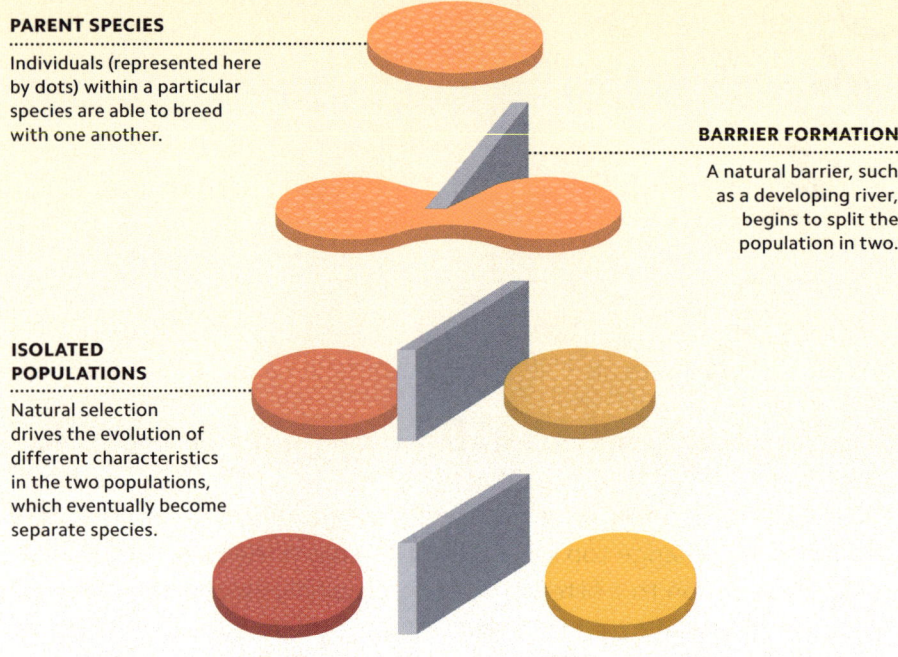

Allopatric speciation
When species arise through separation and isolation
of populations, the process is called allopatric speciation.
In sympatric speciation, new species may arise in one location.

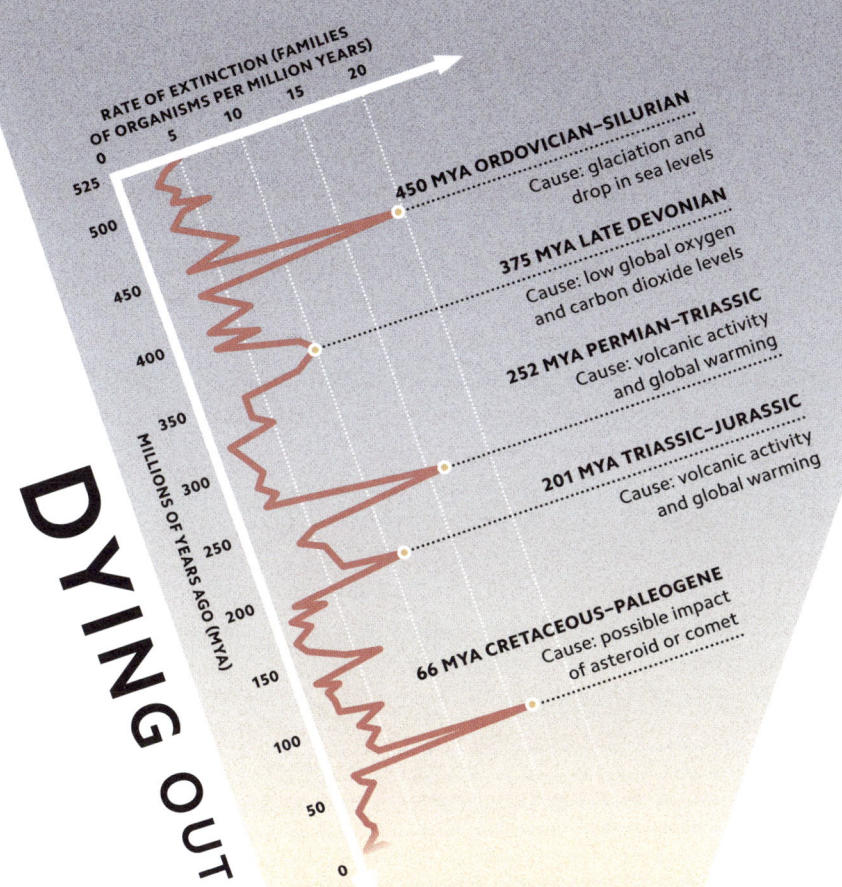

DYING OUT

RATE OF EXTINCTION (FAMILIES OF ORGANISMS PER MILLION YEARS)

0 5 10 15 20

MILLIONS OF YEARS AGO (MYA)

525
500
450
400
350
300
250
200
150
100
50
0

450 MYA ORDOVICIAN–SILURIAN
Cause: glaciation and drop in sea levels

375 MYA LATE DEVONIAN
Cause: low global oxygen and carbon dioxide levels

252 MYA PERMIAN–TRIASSIC
Cause: volcanic activity and global warming

201 MYA TRIASSIC–JURASSIC
Cause: volcanic activity and global warming

66 MYA CRETACEOUS–PALEOGENE
Cause: possible impact of asteroid or comet

Just as species can be formed, they can also disappear. Indeed, more than 99 per cent of all species that have lived on Earth are now extinct. The causes of extinction are diverse: for example, the environment of a species may change too rapidly for it to adapt; the species may lose out in competition to a fitter species; or it may be wiped out by disease. Extinctions may affect one species, or many; the fossil record indicates that there have been five mass extinction episodes that have each eliminated more than 75 per cent of the species alive at that time. Some scientists have proposed that a sixth mass extinction event, caused in large part by human activity, has begun.

Habitat diversity
One ecosystem can contain many individual habitats – for example, a forest has distinct habitats in the canopy, in the soil, on the forest floor, and in tree hollows.

BIOLOGICAL FACTORS

These include food, space, competition, predators, and (for plants) pollinators.

PHYSICAL FACTORS

These include shelter, soil, light, and humidity.

HABITAT

Species can share the same habitat.

Habitats of species can overlap.

LIVING SPACES

The place where an organism or community lives is known as its habitat. It includes all the physical and biological factors needed for the survival of that organism. An ecosystem is a broader concept, encompassing all the organisms in a particular area, and the totality of their interactions with each other as well as the physical environment. Like habitats, ecosystems vary greatly in scale, from tide pools to entire oceans.

ECOSYSTEM

WEBS OF LIFE

Energy and nutrients flow between organisms in an ecosystem. In photosynthesis, plants (producers) absorb energy from sunlight to make energy-dense compounds on which animals feed – either directly, as in herbivores (primary consumers), or indirectly, as in the carnivorous animals (secondary and tertiary consumers) that prey on other species. Decomposers, which may be microbes, plants, fungi, or animals, consume the bodies of dead organisms, freeing nutrients locked up in their tissues. The visual representation of all these relationships in an ecosystem is called a food web.

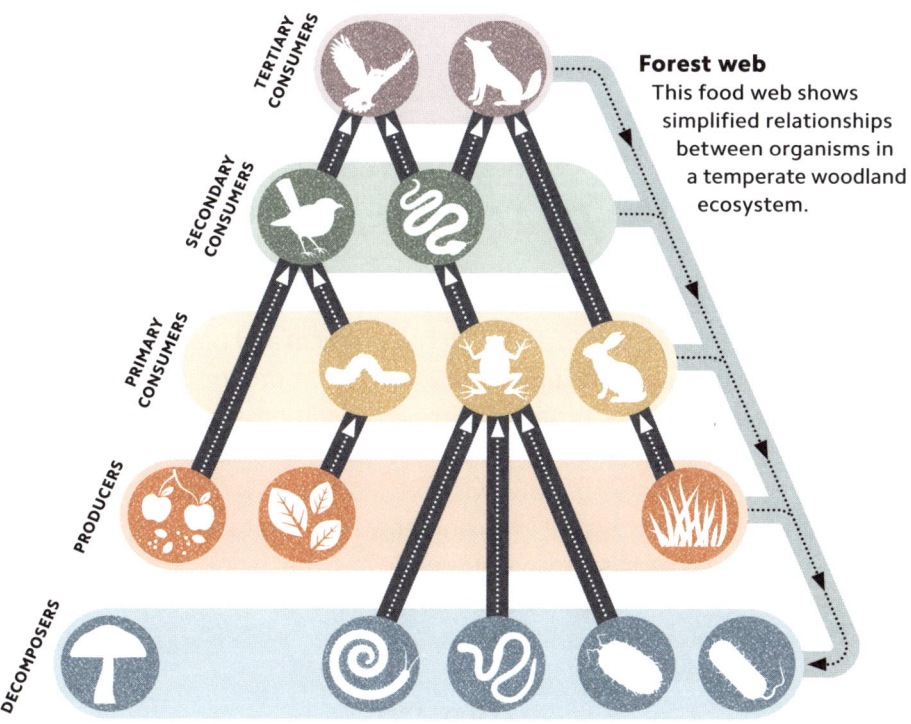

Forest web
This food web shows simplified relationships between organisms in a temperate woodland ecosystem.

TERTIARY CONSUMERS

SECONDARY CONSUMERS

PRIMARY CONSUMERS

PRODUCERS

DECOMPOSERS

EARTH

The structure and makeup of our planet were determined by processes that began 4.6 billion years ago and continue to this day, regularly reminding us of their presence through spectacular volcanic eruptions and destructive earthquakes. We live on a thin, rocky crust that is shaped by processes driven by heat from Earth's interior and radiant energy from the Sun. It is this solar energy that drives the circulation of the oceans and atmosphere that govern our climate and transport water around the globe. The features of Earth's surface – mountains, seas, and glaciers – result from the interaction of the forces from below and above.

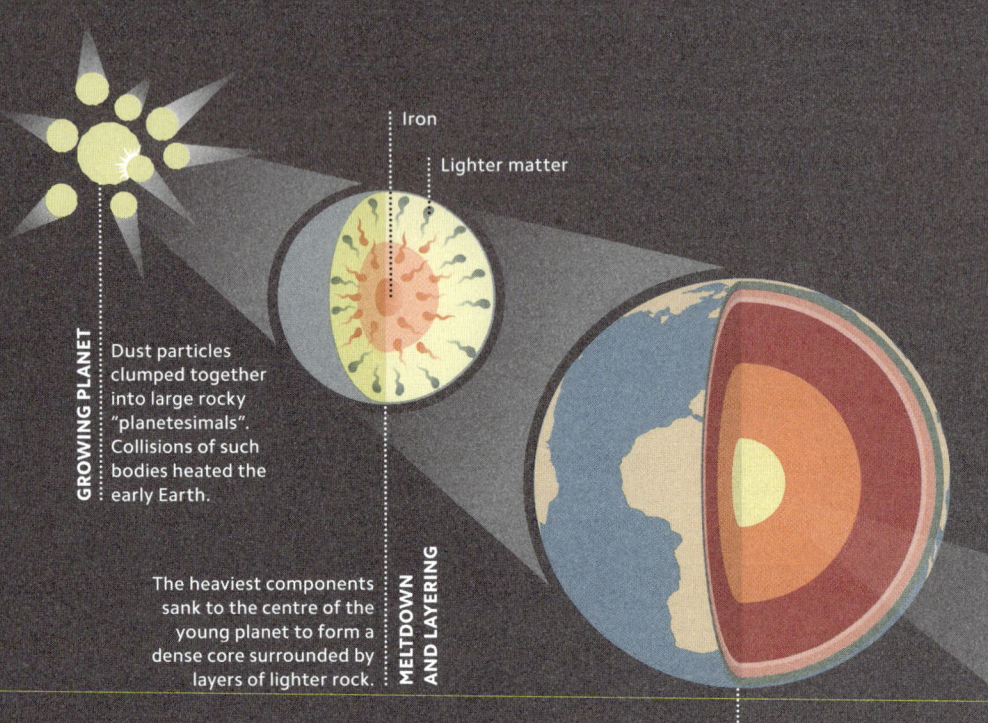

Iron

Lighter matter

GROWING PLANET

Dust particles clumped together into large rocky "planetesimals". Collisions of such bodies heated the early Earth.

MELTDOWN AND LAYERING

The heaviest components sank to the centre of the young planet to form a dense core surrounded by layers of lighter rock.

EARTH TODAY

Earth was transformed from a uniform structure to one with distinct zones.

Formation of Earth
Earth is almost 4.6 billion years old. It was born from the cloud of dust and gas orbiting the Sun.

OUR LAYERED PLANET

Earth has a diameter of 12,756 km (7,926 miles) – a figure approximated by the Ancient Greeks. Despite the lack of direct evidence of our planet's structure, knowledge of its interior has grown enormously over the last hundred years. Advances in equipment able to detect and analyse gravity, magnetism, and seismic (earthquake) waves, along with computer modelling, have revealed Earth's layered structure and processes occurring within the layers. Meteorite studies have helped determine the composition of Earth's deep interior.

Drills have penetrated only the top
12 km (8 miles) of Earth's crust.

INNER CORE
Made from solid metals – mainly iron and nickel – the core is subject to pressures 3 million times those at the surface and temperatures of over 5,000°C (9,000°F).

OUTER CORE
The outer core is made of molten iron and nickel, heated by natural radioactive decay. The flow of this material creates electrical currents that in turn produce Earth's magnetic field.

LITHOSPHERE
The top layer of the upper mantle (the lithospheric mantle) and crust together make up the lithosphere, which "floats" on the asthenosphere below.

ASTHENOSPHERE
This semi-solid layer is weak and ductile and can flow over very long timescales.

INNER CORE

OUTER CORE

MANTLE

ASTHENOSPHERE

LITHOSPHERIC MANTLE

CRUST

This layer of semi-solid rock makes up over 80 per cent of Earth's volume. Its temperature ranges from 1,000 to 3,700°C (1,800 to 6,700°F) at depth.

MANTLE

This outer shell, made of solid rock and minerals, may be 5–80 km (3–50 miles) thick.

CRUST

Where plates meet

Tectonic plates can collide, move apart, or slide past one another; the places where these meetings occur are called convergent or divergent boundaries, or transform fault boundaries.

DIVERGENT BOUNDARY
Where plates move apart, earthquakes occur and molten rock erupts onto the seabed. It then solidifies to create new oceanic crust.

RIDGE PUSH
Slabs of rigid lithosphere slide down and away from the hot raised asthenosphere of the ridge, driving plate motion.

Mid-ocean ridge

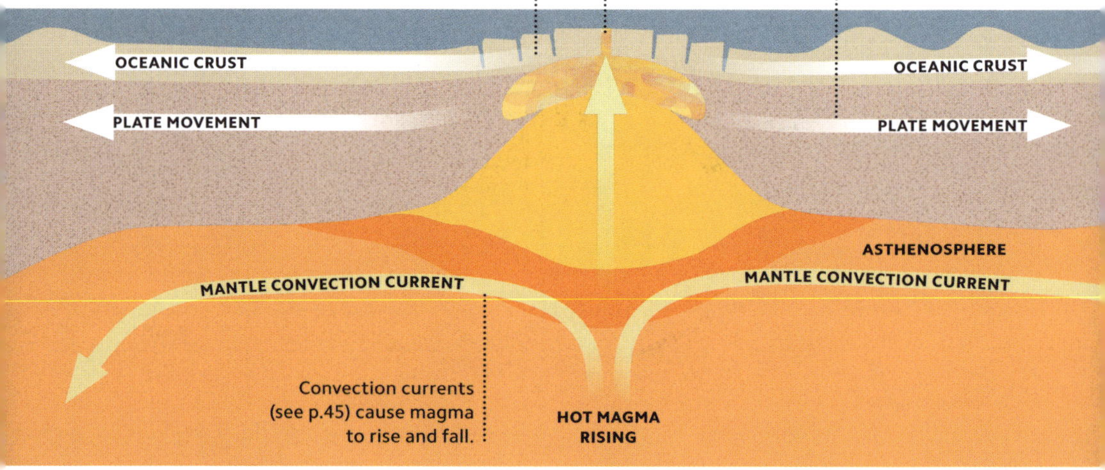

OCEANIC CRUST

PLATE MOVEMENT

OCEANIC CRUST

PLATE MOVEMENT

ASTHENOSPHERE

MANTLE CONVECTION CURRENT

MANTLE CONVECTION CURRENT

Convection currents (see p.45) cause magma to rise and fall.

HOT MAGMA RISING

A jigsaw of plates

The lithosphere is made up of the crust and lithospheric mantle. It is split into seven major plates, as well as many minor ones. The crust under the oceans is usually less than 15 km (9 miles) thick and is made of relatively young rock. Crust under continents is thicker and typically older.

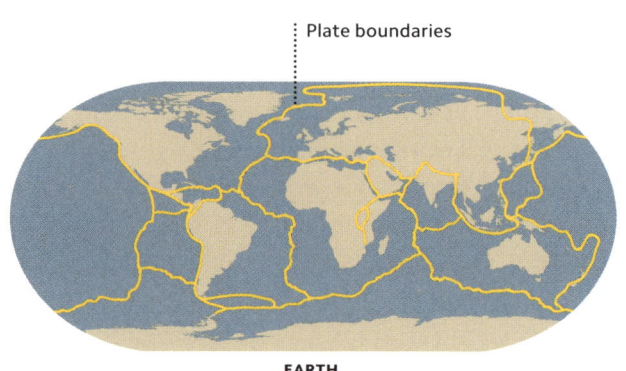

Plate boundaries

EARTH

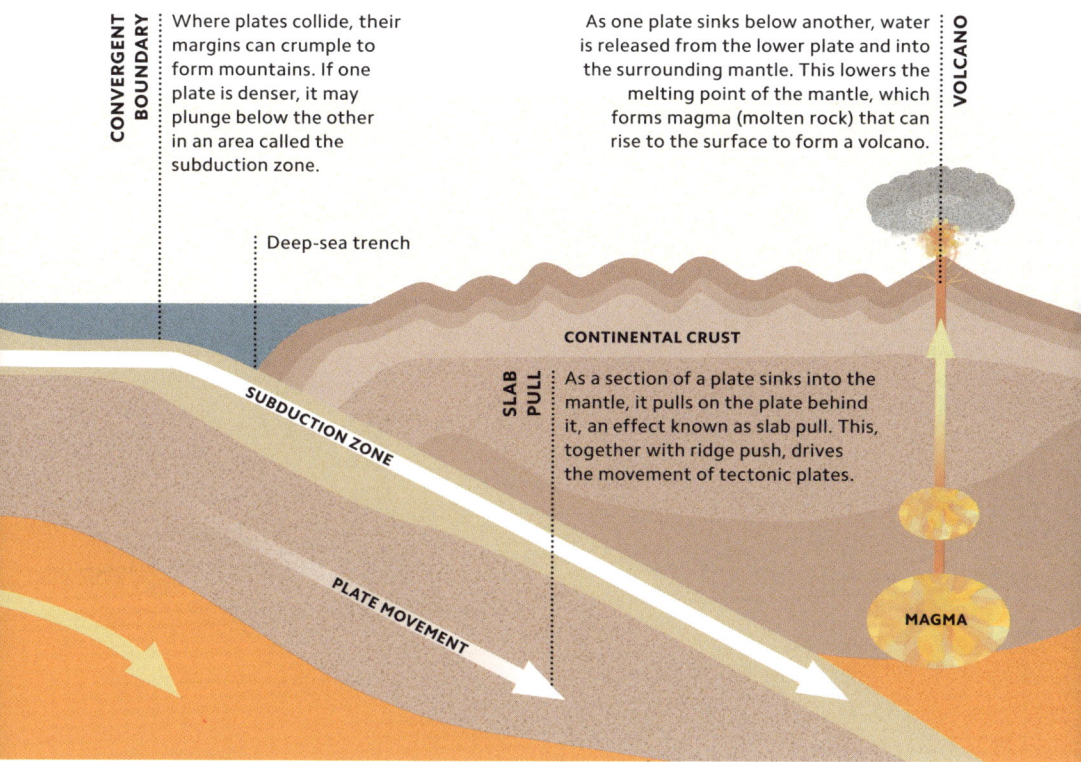

Where plates collide, their margins can crumple to form mountains. If one plate is denser, it may plunge below the other in an area called the subduction zone.

As one plate sinks below another, water is released from the lower plate and into the surrounding mantle. This lowers the melting point of the mantle, which forms magma (molten rock) that can rise to the surface to form a volcano.

VOLCANO

Deep-sea trench

CONTINENTAL CRUST

SUBDUCTION ZONE

SLAB PULL

As a section of a plate sinks into the mantle, it pulls on the plate behind it, an effect known as slab pull. This, together with ridge push, drives the movement of tectonic plates.

PLATE MOVEMENT

MAGMA

SHIFTING SURFACE

The theory of plate tectonics revolutionized Earth science in the 1960s by unifying the explanations for phenomena such as volcanoes, earthquakes, mountain formation, and the distribution of different rock types in Earth's crust. Plate tectonics proposes that Earth's outer shell – the solid but brittle lithosphere – is broken up into plates that carry the continents. These "float" on the asthenosphere and can move relative to one another. Their movement over billions of years has governed the shape and position of Earth's continents.

Volcanoes are structures that form where magma (molten rock) erupts at Earth's surface. They often consist of a cone with a crater (depression) at the top. Most form at the boundaries of tectonic plates (see pp.130–31) but some arise in the middle of plates above very hot plumes of magma. Around three-quarters of Earth's active volcanoes are located in the "Ring of Fire" around the edges of the Pacific Ocean, where tectonic activity is very high.

ACTIVE EARTH

ERUPTION

A volcano erupts when hot expanding magma rises to the surface through pipes or fissures. If the magma is viscous (sticky) and contains a lot of dissolved gas (which is released as magma rises), the eruption is likely to be violent.

VENT

VOLCANIC CLOUD

Airborne fragments of rock and lava are known as pyroclasts.

PYROCLASTIC FLOW

A mixture of hot lava blocks, ash, and hot gas rushes down the volcano.

DIKES

Magma can solidify within the volcano in dikes.

FISSURES

Magma can leak out through cracks called fissures.

Pressurized magma from the mantle rises up to fill and expand cracks in the rock above, forming a large magma chamber.

CONTINENTAL CRUST

LITHOSPHERIC MANTLE

MAGMA CHAMBER

ASTHENOSPHERE

A magnitude 8 earthquake releases as much energy as detonating 15 million tons of TNT.

Like volcanoes, earthquakes can occur where two tectonic plates collide or move apart (pp.130–31).

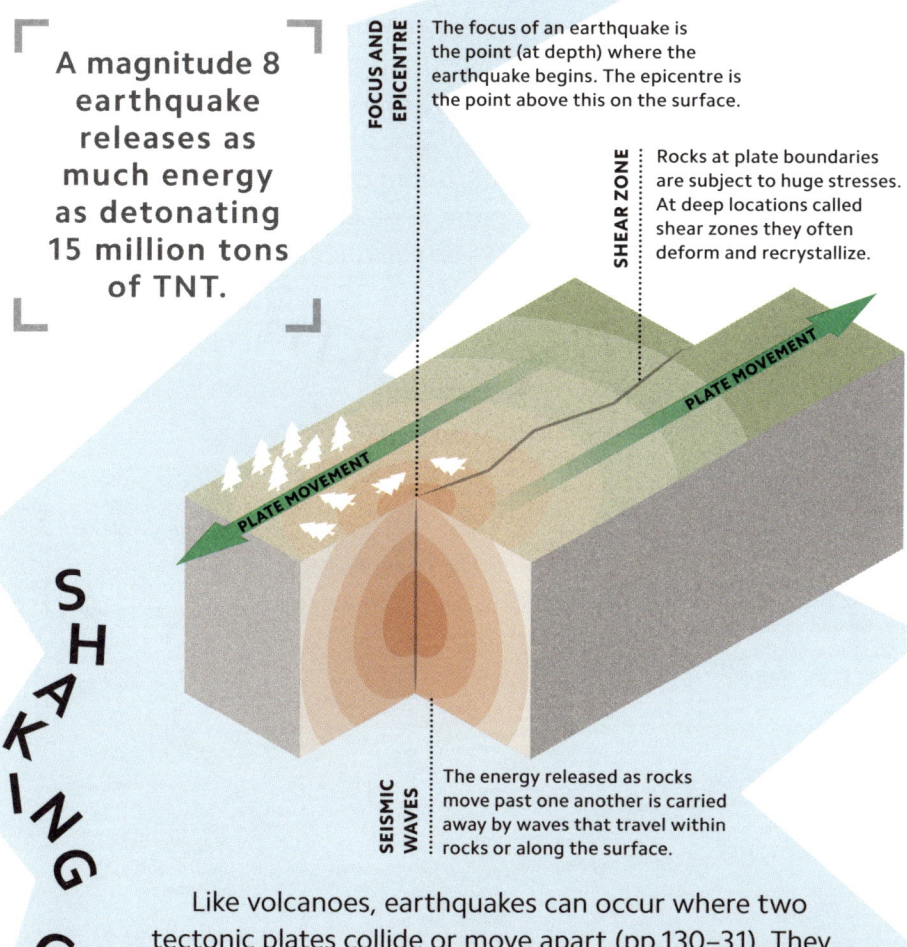

PLATE MOVEMENT

PLATE MOVEMENT

FOCUS AND EPICENTRE

The focus of an earthquake is the point (at depth) where the earthquake begins. The epicentre is the point above this on the surface.

SHEAR ZONE

Rocks at plate boundaries are subject to huge stresses. At deep locations called shear zones they often deform and recrystallize.

SEISMIC WAVES

The energy released as rocks move past one another is carried away by waves that travel within rocks or along the surface.

SHAKING GROUND

Like volcanoes, earthquakes can occur where two tectonic plates collide or move apart (pp.130–31). They also occur at transform fault boundaries – where two plates are moving past one another. Friction between two plates makes them "stick" until the energy stored in this tension is suddenly released. Shock waves in the rock carry that energy away from the focus of the earthquake. The amount of energy released is quantified using the moment magnitude scale, which is similar to the older Richter scale.

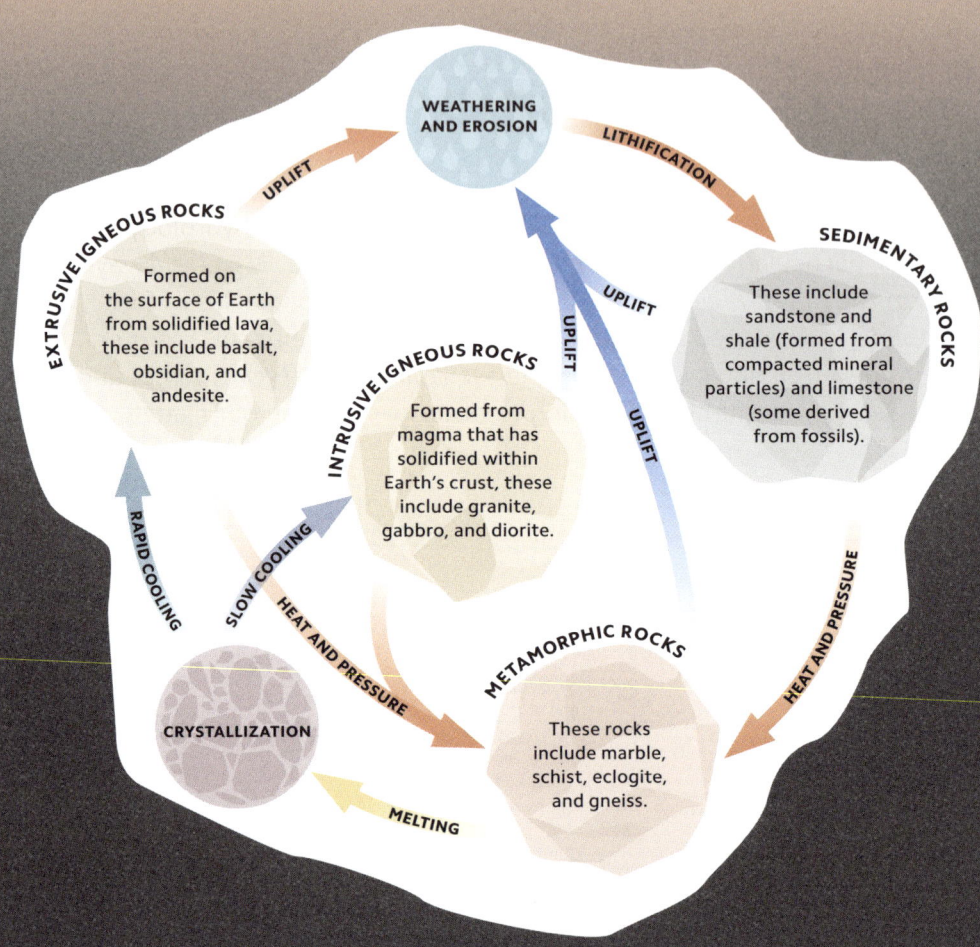

WEATHERING AND EROSION

UPLIFT

LITHIFICATION

EXTRUSIVE IGNEOUS ROCKS

Formed on the surface of Earth from solidified lava, these include basalt, obsidian, and andesite.

SEDIMENTARY ROCKS

These include sandstone and shale (formed from compacted mineral particles) and limestone (some derived from fossils).

INTRUSIVE IGNEOUS ROCKS

Formed from magma that has solidified within Earth's crust, these include granite, gabbro, and diorite.

UPLIFT

UPLIFT

UPLIFT

RAPID COOLING

SLOW COOLING

HEAT AND PRESSURE

METAMORPHIC ROCKS

These rocks include marble, schist, eclogite, and gneiss.

HEAT AND PRESSURE

CRYSTALLIZATION

MELTING

TOUGH CHANGES

Most rocks in Earth's crust are made of a mixture of minerals. Many are rich in silica, but others vary in composition depending on where and how they formed. Most are igneous, formed when magma solidifies (see pp.130 –31); some are metamorphic, formed when existing rock is transformed by heat and pressure; and others are sedimentary, mostly formed at Earth's surface when mineral or organically derived particles settle and become cemented together.

LIQUID CIRCULATION

Water is essential to life. Numerous intertwined processes determine its distribution and quality, and their study – known as hydrology – is vital, particularly in these times of changing climate. Water is always on the move: it evaporates from land and seas, condenses to form clouds, is deposited as liquid or ice, and percolates into soil and rocks. This never-ending cycle affects the availability of water resources and is a key factor in determining weather patterns on our planet.

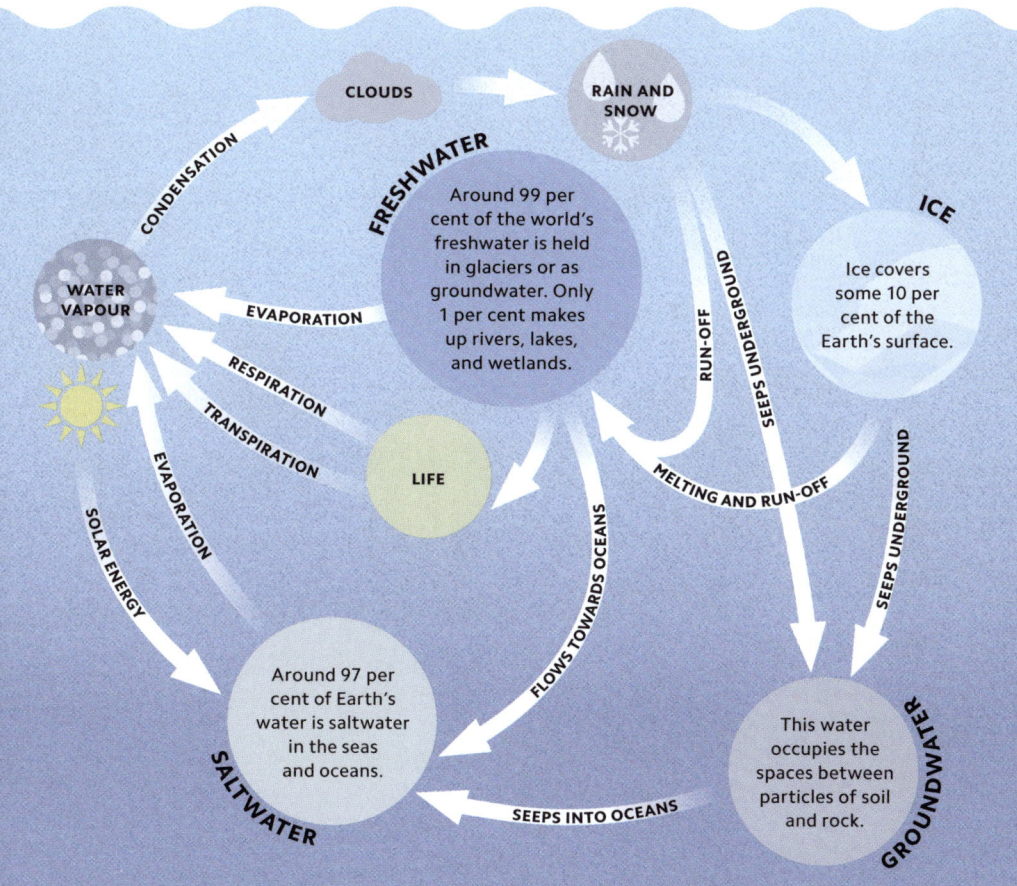

CLOUDS

RAIN AND SNOW

CONDENSATION

FRESHWATER

Around 99 per cent of the world's freshwater is held in glaciers or as groundwater. Only 1 per cent makes up rivers, lakes, and wetlands.

ICE

Ice covers some 10 per cent of the Earth's surface.

WATER VAPOUR

EVAPORATION

RESPIRATION

TRANSPIRATION

RUN-OFF

SEEPS UNDERGROUND

LIFE

MELTING AND RUN-OFF

SEEPS UNDERGROUND

SOLAR ENERGY

EVAPORATION

FLOWS TOWARDS OCEANS

Around 97 per cent of Earth's water is saltwater in the seas and oceans.

This water occupies the spaces between particles of soil and rock.

GROUNDWATER

SALTWATER

SEEPS INTO OCEANS

AN OCEAN OF AIR

The thin layer of mixed gases around Earth makes our planet habitable. As well as being a source of oxygen for respiration and of carbon dioxide for photosynthesis, it screens out damaging radiation while retaining heat and moisture. At altitudes below around 100 km (60 miles), the composition of the atmosphere is relatively constant – predominantly nitrogen and oxygen (see p.25) – though its density and pressure decrease with elevation (see below). At higher altitudes, gases such as ozone, helium, and hydrogen predominate.

Layers of the atmosphere
Meteorologists recognize five layers of the atmosphere. These are distinguished by the temperature gradients within them, rather than their height or composition.

At a height of 18 km (11 miles), known as the Armstrong limit, atmospheric pressure is so low that human blood boils.

EXOSPHERE
600–10,000 KM (370–6,000 MILES)

THERMOSPHERE
80–600 KM (50–370 MILES)

SATELLITES

Here, the atmosphere gives way to space. The exosphere is fully exposed to the solar wind – a stream of high-speed particles expelled from the Sun.

Air is so thin at this altitude that it can transfer very little energy, so the thermosphere is extremely cold. It contains the ionosphere – a region where solar energy causes gases to split into ions and free electrons.

EARTH'S ATMOSPHERE

Atmospheric pressure

The air around us exerts a pressure that depends on its depth and density. Both are at their greatest at Earth's surface; some 50 per cent of the molecules of the atmosphere are within 7 km (4 miles) of the surface. The pressure of one Earth atmosphere is equivalent to 10 tonnes of weight applied on one square metre of area.

Pressure at Everest's summit is one-third that at sea level. At 100 km (60 miles) high, pressure is effectively zero.

DECREASING AIR PRESSURE

MT EVEREST, 8,849 M (29,032 FT)

SEA LEVEL

A pressure of one atmosphere is air pressure measured at sea level

---

MESOPHERE
50–80 KM (30–50 MILES)

STRATOSPHERE
16–50 KM (10–30 MILES)

TROPOSPHERE
0–16 KM (0–10 MILES)

METEORS

AIRCRAFT

TEMPERATURE

OZONE LAYER

WEATHER BALLOON

TEMPERATURE

| -100°C | -80°C | -60°C | -40°C | -20°C | 0°C | 20°C | 40°C |
| -148°F | -112°F | -76°F | -40°F | -4°F | 32 F | 68 F | 104 F |

In this layer, there are relatively few gas molecules to absorb the Sun's energy, so the mesosphere is warmer at its base (where it gets some heat from the stratosphere) than at its top.

Here, temperature increases with altitude as air is warmed directly by the Sun. With warm air at the top and cold at the bottom, the stratosphere is relatively stable. A layer of ozone gas, which absorbs ultraviolet radiation that is harmful to life, is found in the stratosphere.

When solar energy hits the surface, air at low levels of the troposphere is warmed. This warm air rises up through cooler air in the troposphere, before it cools and falls. This movement and mixing of air helps drive our planet's weather.

Air in the polar circulation sinks at the highest latitudes. This circulation is weak because it is driven by little solar energy.

POLAR CELLS

Warm air ascends where it meets cool air from the poles, creating another circulation loop called a Ferrel cell. It generates westerly winds.

FERREL CELLS

Hot air rising at the equator creates a circulation loop called a Hadley cell. Air flowing back to the equator is deflected to the east by the rotation of Earth.

HADLEY CELLS

POLAR REGION
ARCTIC CIRCLE

WESTERLIES

TROPIC OF CANCER

TRADE WINDS

EQUATOR

EARTH'S AXIS

Global winds

At the equator, hot air rises high in the troposphere and moves north or south before cooling, falling, and flowing back towards the equator. This flow, combined with Earth's rotation, produces easterly trade winds. Similar circulation cells drive wind patterns further to the north and south.

FLOWING AIR

The Sun warms Earth's surface unevenly. Features such as rocks, water, and vegetation absorb heat at different rates, so the air above them heats up patchily. As warm air rises, cooler air flows in from the sides to take its place. This flow is wind. It can be generated at a variety of scales, from local to global. On a small scale, wind can be generated as air above land heatsmore rapidly than does air over adjacent sea. As this air rises, cooler air from the sea takes its place, producing a sea breeze.

CHANGE IS IN THE AIR

Movements of air masses of different temperature and humidity translate into changes in the weather, which meteorologists describe in terms of fronts and shifts in pressure. A front is simply the boundary between two different air masses, such as a mass of warm, moist air and a mass of cold, drier air. Fronts are propelled by winds, and their arrival signals changes in temperature, wind, and precipitation.

ADVANCING WARM FRONT

WARM AIR RISES

CLOUD BAND

On reaching colder – and thus denser – air, a warm air mass is forced upwards, causing pressure to fall.

MASS OF WARM AIR

A warm front is the front edge of a mass of advancing warm, humid air.

Eventually, the warm air fully displaces the cold; pressure increases and temperature rises.

The air cools. The water it holds condenses, falling as precipitation.

WARM FRONT

MASS OF COLD AIR

COLD AIR

CARBON PATHWAYS

Carbon is essential to life. Like water, it is a finite resource and is constantly cycled through living and non-living things. More than 99 per cent of the carbon on our planet is locked up in sediments or rocks, such as limestone, and deposits that include fossil fuels. The next largest store is the oceans (in the form of dissolved carbon dioxide), followed by soil (mainly as carbonate minerals), the atmosphere (in the form of carbon dioxide gas), and then living organisms. Carbon moves between these stores in many different ways, and it can reside within in them for very long or very short periods.

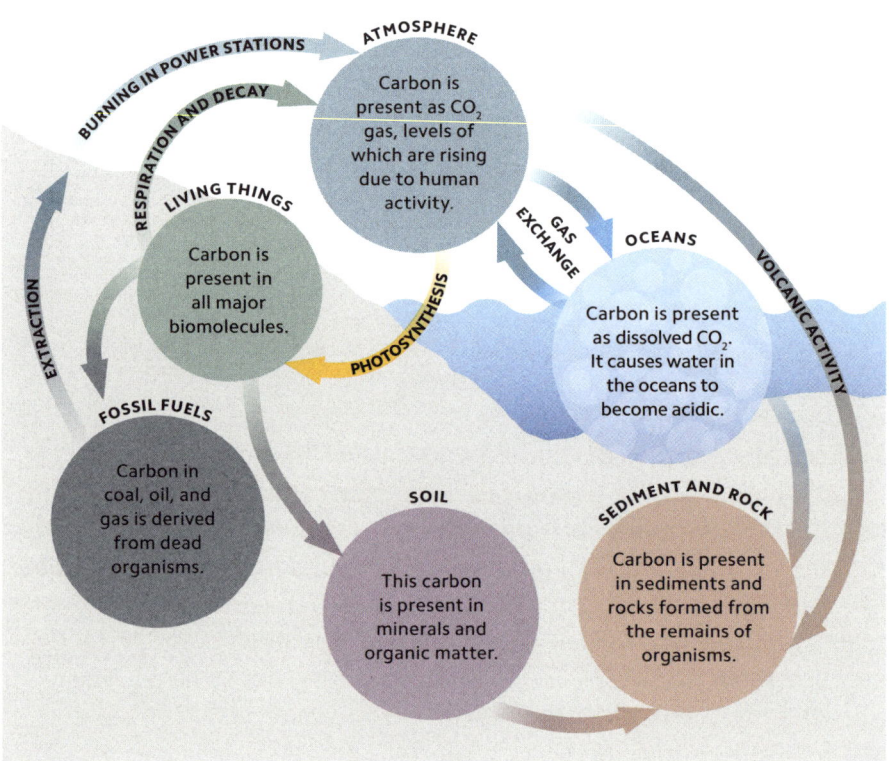

ATMOSPHERE
Carbon is present as CO_2 gas, levels of which are rising due to human activity.

BURNING IN POWER STATIONS

RESPIRATION AND DECAY

LIVING THINGS
Carbon is present in all major biomolecules.

EXTRACTION

GAS EXCHANGE

OCEANS
Carbon is present as dissolved CO_2. It causes water in the oceans to become acidic.

VOLCANIC ACTIVITY

PHOTOSYNTHESIS

FOSSIL FUELS
Carbon in coal, oil, and gas is derived from dead organisms.

SOIL
This carbon is present in minerals and organic matter.

SEDIMENT AND ROCK
Carbon is present in sediments and rocks formed from the remains of organisms.

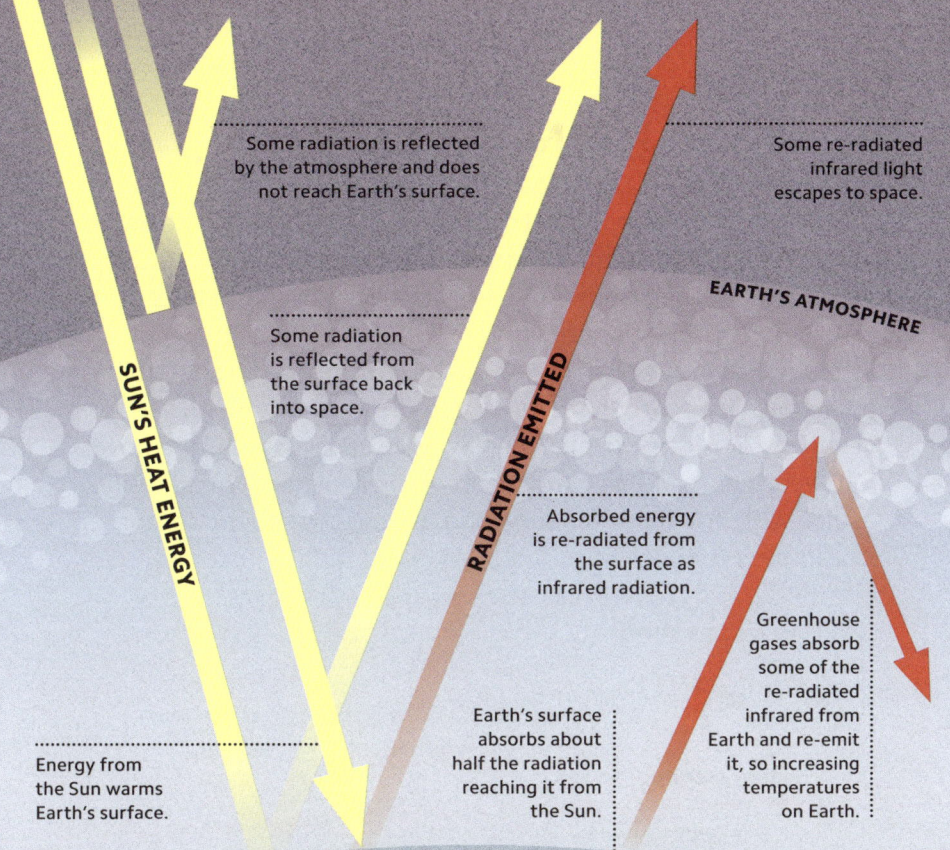

Some radiation is reflected
by the atmosphere and does
not reach Earth's surface.

Some re-radiated
infrared light
escapes to space.

Some radiation
is reflected from
the surface back
into space.

EARTH'S ATMOSPHERE

SUN'S HEAT ENERGY

RADIATION EMITTED

Absorbed energy
is re-radiated from
the surface as
infrared radiation.

Greenhouse
gases absorb
some of the
re-radiated
infrared from
Earth and re-emit
it, so increasing
temperatures
on Earth.

Energy from
the Sun warms
Earth's surface.

Earth's surface
absorbs about
half the radiation
reaching it from
the Sun.

TEMPERATURE CONTROL

Earth's atmosphere is a thin layer of gases that plays a vital role in
regulating surface temperatures. It reflects some light from the Sun
back into space, moderating daytime temperatures, and – like the glass
of a greenhouse – traps some of the heat that would otherwise radiate
from Earth back out into space, preventing extreme cold at night. Some
gases, such as carbon dioxide, methane, and water, are very effective
at retaining this heat, and so are called greenhouse gases. Their
presence is necessary for Earth, but global temperatures are
very sensitive to their concentration in the atmosphere.

MERCURY RISING

Earth's climate undergoes natural variations, mostly caused by small irregularities in our planet's orbit around the Sun. However, since the mid-19th century, the average temperature on Earth has been rising at an unprecedented rate, increasing by about 1.1 °C (2.0 °F) since 1880, and projected to rise by 2.0–4.0 °C (3.6–7.2 °F) by 2100. The evidence that this increase is caused by human activity – principally the burning of fossil fuels by industrialized countries – is now overwhelming. The release of combustion products, notably carbon dioxide, has affected the carbon cycle (see p.140), causing excess greenhouse gases to persist in our atmosphere.

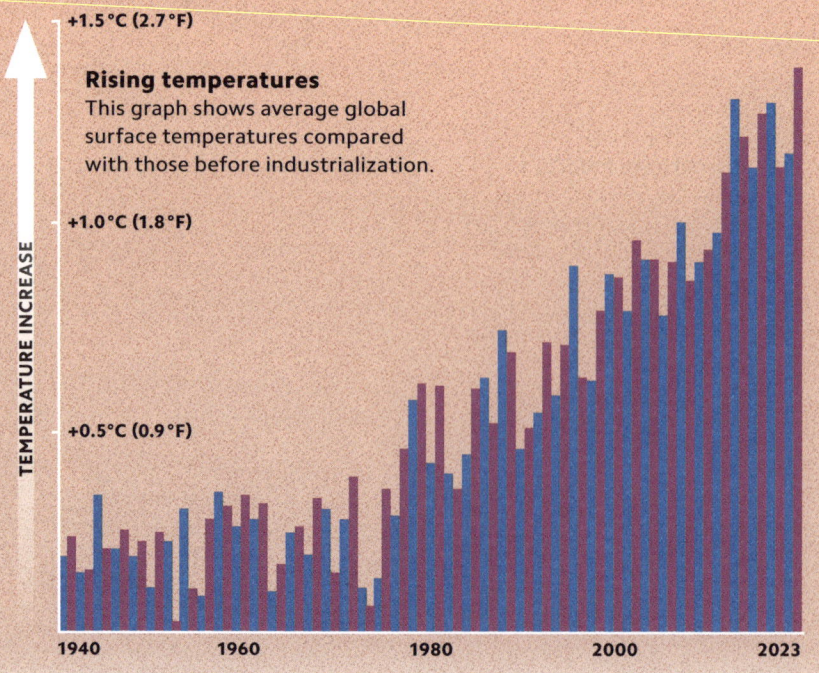

+1.5 °C (2.7 °F)

Rising temperatures
This graph shows average global surface temperatures compared with those before industrialization.

+1.0 °C (1.8 °F)

TEMPERATURE INCREASE

+0.5 °C (0.9 °F)

1940   1960   1980   2000   2023

The effects of global warming are far reaching. The additional heat energy retained within Earth's atmosphere by greenhouse gases causes shifts in weather patterns and seasonality – effects felt by all living things. What were once rare extreme weather events involving heat or rainfall have become more common, causing ever more destructive floods, wildfires, famines, droughts, and hurricanes. Global warming also has direct effects on sea levels through the faster melting of glaciers and ice caps, and the thermal expansion of water.

CLIMATE WARNING

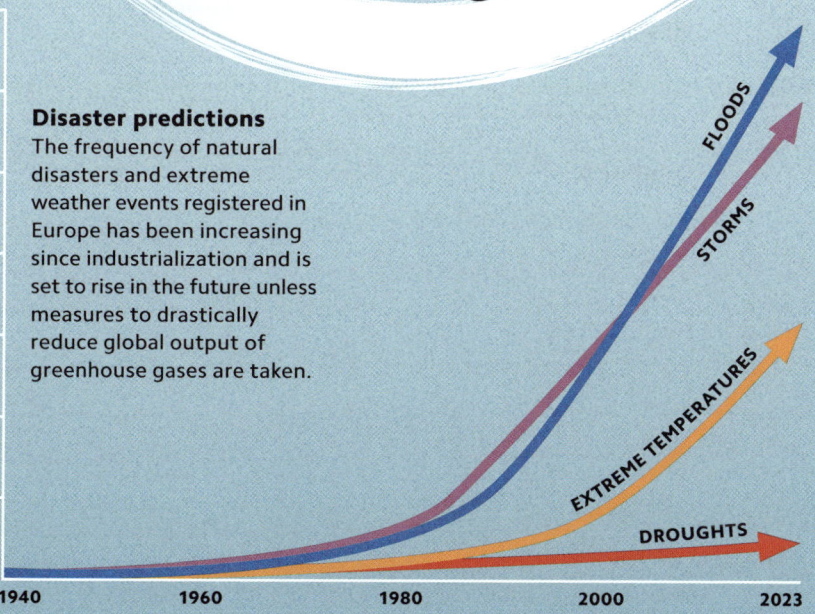

Disaster predictions
The frequency of natural disasters and extreme weather events registered in Europe has been increasing since industrialization and is set to rise in the future unless measures to drastically reduce global output of greenhouse gases are taken.

NUMBER OF NATURAL DISASTERS PER YEAR

600

400

200

FLOODS

STORMS

EXTREME TEMPERATURES

DROUGHTS

1940   1960   1980   2000   2023

ASTRO

N O M Y

Human curiosity about the Universe has remained undimmed for millennia. Our understanding of the origins and nature of planets and stars has grown enormously in recent years, driven by improving techniques and technologies used to view and analyse astronomical phenomena. Developments in mathematics and physics have gifted astronomers new theoretical frameworks to explain once unimaginable phenomena, such as black holes and dark matter. They have also revealed that the properties of stars and galaxies are determined by processes that occur at the scale of atoms and their constituent particles.

Light limits

Light moves at 300,000 km (186,000 miles) per second through the vacuum of space. In one year, it travels about 9.5 trillion km (5.9 trillion miles). This distance is called a light-year. From Earth, we can see only as far as light has been able to travel in the time since the Big Bang (see p.154). This distance is 13.8 billion light-years.

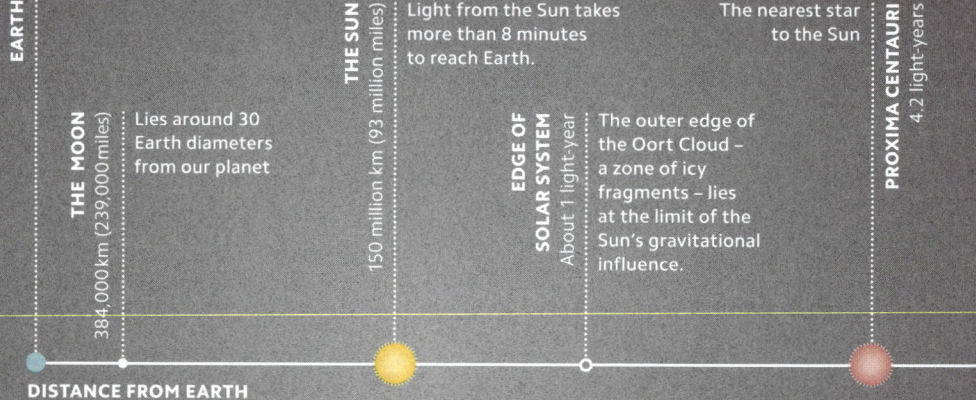

EARTH

THE MOON
384,000 km (239,000 miles)

Lies around 30 Earth diameters from our planet

THE SUN
150 million km (93 million miles)

Light from the Sun takes more than 8 minutes to reach Earth.

EDGE OF SOLAR SYSTEM
About 1 light-year

The outer edge of the Oort Cloud – a zone of icy fragments – lies at the limit of the Sun's gravitational influence.

The nearest star to the Sun

PROXIMA CENTAURI
4.2 light-years

DISTANCE FROM EARTH

EVERYTHING OUT THERE

The size of the Universe is unknown. Instruments have detected galaxies more than 13.4 billion light-years away, but are limited in how far they can probe by the speed of light (see above). Some cosmologists have calculated a diameter for the Universe of at least 23 trillion light-years; others propose that it is infinite. The part that we can potentially see, known as the observable Universe, includes the hundreds of billions of stars of our galaxy, the Milky Way, and hundreds of billions of galaxies beyond. These objects are made from matter, which – astonishingly – may account for only a small proportion of the total mass of the Universe (see opposite).

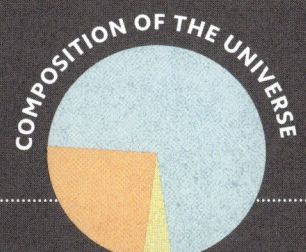

COMPOSITION OF THE UNIVERSE

DARK MATTER

This makes up about 24 per cent of the Universe. It may be composed of subatomic particles that do not interact with electromagnetic radiation.

DARK ENERGY

This little-understood form of energy makes up over 70 per cent of the Universe. It works to push the Universe outwards, counteracting gravity.

ORDINARY MATTER

This makes up less than 5 per cent of the Universe, accounting for all visible objects.

EDGE OF LOCAL GROUP
About 5 million light-years

The group of galaxies to which the Milky Way belongs

THE UNIVERSE BEYOND
Possibly infinite

The size of the Universe is unknown.

ANDROMEDA GALAXY
2.5 million light-years

The closest galaxy to the Milky Way

FURTHEST KNOWN GALAXY
13.4 billion light-years

Space telescopes have detected distant galaxies close to the edge of the observable Universe.

EDGE OF THE OBSERVABLE UNIVERSE
13.8 billion light-years

The limit imposed by the speed of light

"The more the Universe seems comprehensible, the more it also seems pointless."
Steven Weinberg

Terrestrial planets
The four inner plants –
Mercury, Venus, Earth,
and Mars – all have
solid rocky surfaces.

MERCURY (0.39 AU)

Mercury orbits
the Sun in 88 days.
It is a small planet,
little bigger than
our Moon.

VENUS (0.72 AU)

Similar in size to Earth, Venus
has a thick atmosphere that
makes it the hottest planet
in the Solar System. It orbits
the Sun every 225 days.

EARTH (1.0 AU)

THE SUN

The Sun contains most
of the mass of the Solar
System. Planets orbit the
Sun in roughly circular paths.

Earth is just the right
distance from the Sun to
support liquid water
on its surface.

MARS (1.52 AU)

About half the diameter
of Earth, Mars has a
thin atmosphere that
leads to low surface
temperatures. It orbits
the Sun every 687 days.

A question of scale
The Solar System is around 30
trillion km (18.6 trillion miles) in
diameter. Distances within it are
usually measured in astronomical
units (AU), where 1 AU is the
distance between Earth and the
Sun, equivalent to 150 million km
(93 million miles).

ASTEROID BELT
(2.2 – 3.2 AU)

This region contains
orbiting rocky objects,
the largest of which is the
dwarf planet Ceres, some
950 km (590 miles) across.

OUR LOCAL STAR SYSTEM

The Solar System consists of all the objects that are in the Sun's orbit, including eight planets and their moons, at least five dwarf planets, and smaller bodies, such as comets, asteroids, rocks, and icy fragments. Most of it, however, is empty space. It was formed around 4,600 million years ago when a dense cloud of dust and gas collapsed, perhaps as a result of shockwaves from a supernova (exploding star). The Solar System is located on the Orion Arm, a minor spiral arm of our Milky Way galaxy. It orbits its centre at a speed of around 828,000 kph (515,000 mph).

About 11 times the diameter of Earth, Jupiter has around 80 moons. A single orbit of the Sun takes 12 years.

JUPITER (5.20 AU)

SATURN (9.57 AU)

Saturn is orbited by multiple moons and circled rings of icy particles. Its passage around the Sun takes 29.5 years.

KUIPER BELT (30–50 AU)

This is a region of rock, dust and ice. Far beyond this, between 2,000 and about 200,000 AU from the Sun, is a region of icy particles called the Oort Cloud.

URANUS (19.16 AU)

Uranus has the coldest surface of any planet in the Solar System. It orbits the Sun in 84 years.

NEPTUNE (30.18 AU)

This planet's blue colour comes from methane in its atmosphere. Its orbit around the Sun takes 165 years.

PLUTO (39.48 AU)

Distant Pluto was demoted from a planet to dwarf planet status in 2006. A single orbit of the Sun takes 248 years.

Giant planets
These include the gas giants Jupiter and Saturn and the icy planets Uranus and Neptune.

BALLS OF FUSION

A star is a giant ball of hot, glowing gas held together by gravity. The pressure and temperature within its core are high enough to sustain the process of nuclear fusion (see p.84), which liberates vast amounts of energy, including light and other forms of electromagnetic radiation (see pp.70–71). Stars have life cycles: they are born (see p.152) and persist for many millions of years before they die. For about 90 per cent of its life, a star shines by fusing hydrogen into helium.

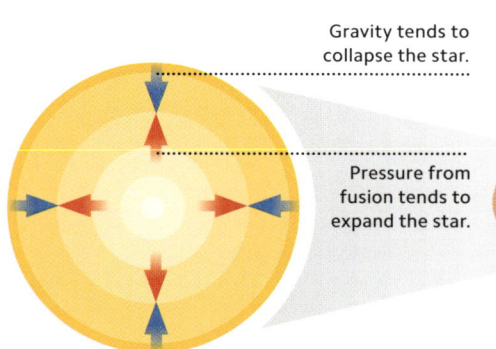

Gravity tends to collapse the star.

Pressure from fusion tends to expand the star.

Gravity and fusion
Stars are stable when the force of gravity crushing them is balanced by the outward pressure from fusion.

YOUNG STAR
Young stars may be surrounded by a disc of gas and dust that can develop into a planetary system.

The largest known star, WOH G64, has the diameter of 1,540 Suns.

Star lives
In the middle of their lives, stars vary greatly in colour, temperature, and mass, from about 0.1 times the mass of the Sun to about 200 times. The later stages of a star's life, and its total lifespan, depend critically on its mass.

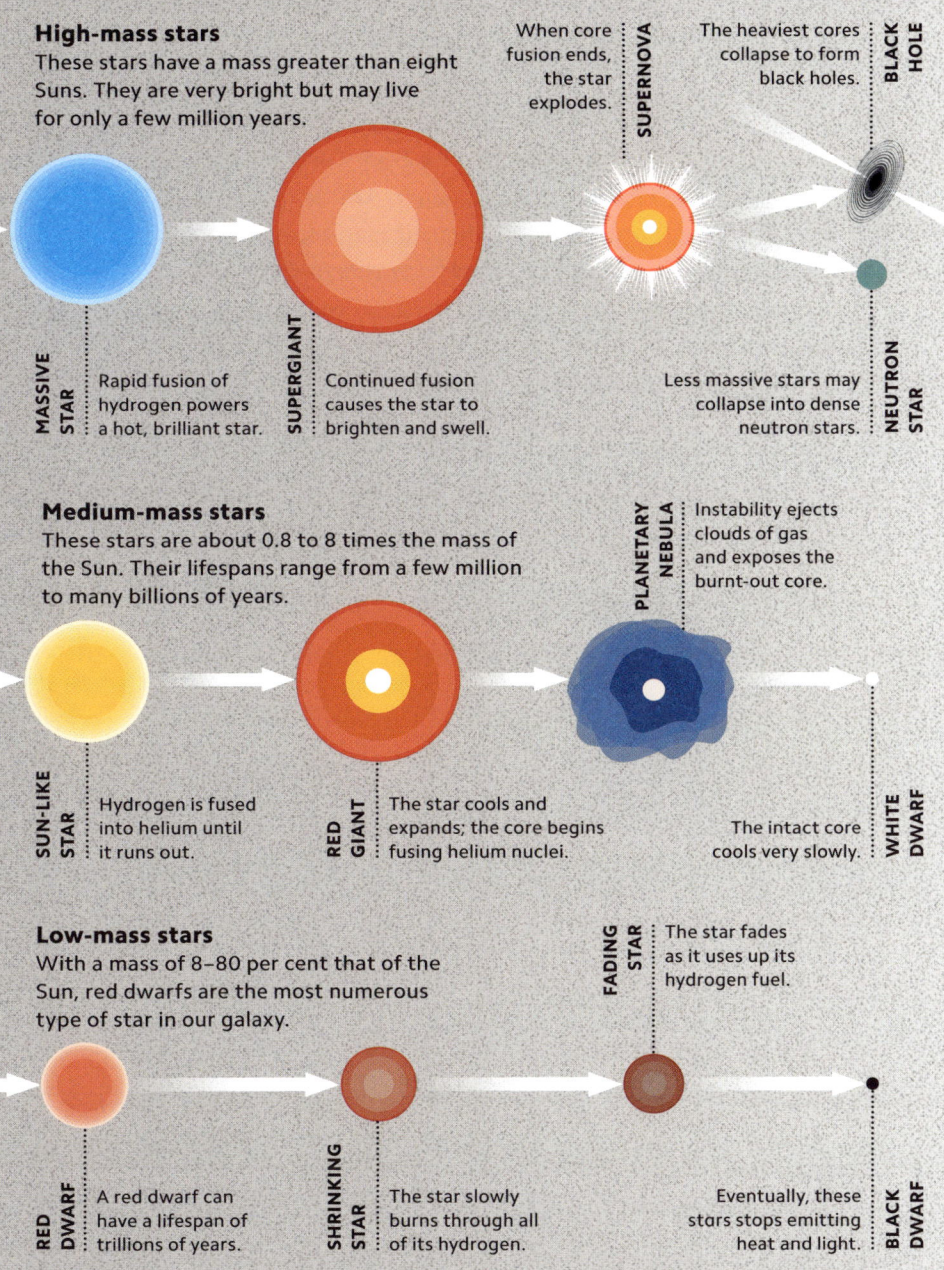

High-mass stars

These stars have a mass greater than eight Suns. They are very bright but may live for only a few million years.

MASSIVE STAR
Rapid fusion of hydrogen powers a hot, brilliant star.

SUPERGIANT
Continued fusion causes the star to brighten and swell.

When core fusion ends, the star explodes.

SUPERNOVA

The heaviest cores collapse to form black holes.

BLACK HOLE

Less massive stars may collapse into dense neutron stars.

NEUTRON STAR

Medium-mass stars

These stars are about 0.8 to 8 times the mass of the Sun. Their lifespans range from a few million to many billions of years.

SUN-LIKE STAR
Hydrogen is fused into helium until it runs out.

RED GIANT
The star cools and expands; the core begins fusing helium nuclei.

PLANETARY NEBULA
Instability ejects clouds of gas and exposes the burnt-out core.

The intact core cools very slowly.

WHITE DWARF

Low-mass stars

With a mass of 8–80 per cent that of the Sun, red dwarfs are the most numerous type of star in our galaxy.

RED DWARF
A red dwarf can have a lifespan of trillions of years.

SHRINKING STAR
The star slowly burns through all of its hydrogen.

FADING STAR
The star fades as it uses up its hydrogen fuel.

Eventually, these stars stops emitting heat and light.

BLACK DWARF

A STAR IS BORN

Stars form in vast clouds of dust and gas called nebulae that may be millions of times the mass of our Sun. Turbulence within these extremely cold clouds causes matter to pile up in some areas more than others. Gravitational forces then begin to pull the denser areas together, forming a core in which temperature and pressure soon rise. Over tens of thousands of years, the core contracts, and the whole system begins to rotate, attracting more matter that gathers around the "equator" of the developing star (protostar), forming a disc, while gas is ejected at the poles. Material from the disc is added to the core, which eventually becomes sufficiently hot and dense to initiate nuclear fusion.

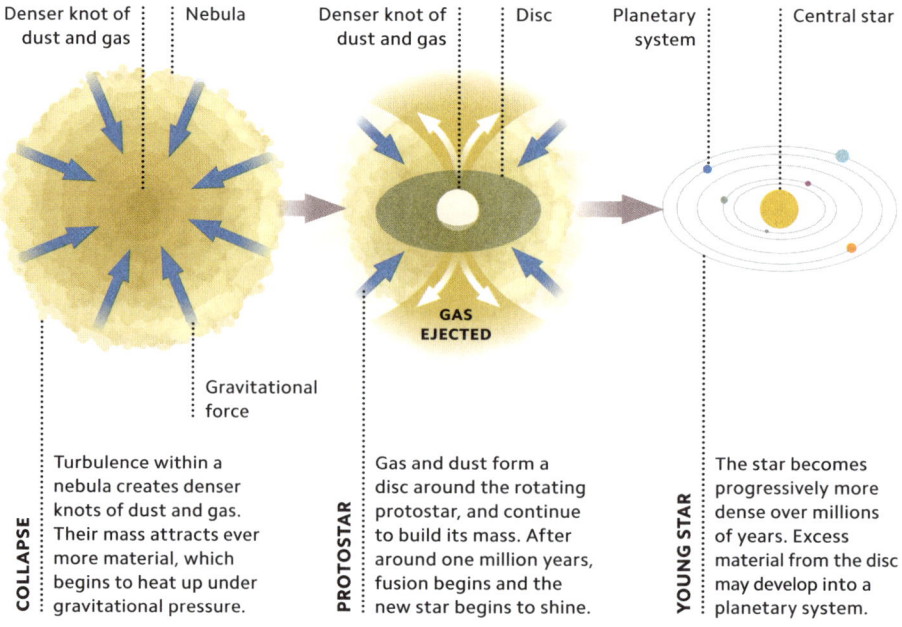

Denser knot of dust and gas | Nebula | Denser knot of dust and gas | Disc | Planetary system | Central star

Gravitational force

GAS EJECTED

COLLAPSE
Turbulence within a nebula creates denser knots of dust and gas. Their mass attracts ever more material, which begins to heat up under gravitational pressure.

PROTOSTAR
Gas and dust form a disc around the rotating protostar, and continue to build its mass. After around one million years, fusion begins and the new star begins to shine.

YOUNG STAR
The star becomes progressively more dense over millions of years. Excess material from the disc may develop into a planetary system.

Anatomy of a galaxy

Galaxies range from hundreds to hundreds of thousands of light-years across. They also come in various shapes. The Milky Way is a flattened spiral shape 100,000–120,000 light-years across and contains up to 400 billion stars.

The Sun is located on a minor spiral arm of the galaxy, some 27,000 light years from its centre.

SUN

THICK DISC

Located at the centre of the Milky Way, this region is densely packed with stars.

NUCLEAR BULGE

THIN DISC

GALACTIC DISC

This is an accumulation of stars, gas, and dust that orbits the galactic centre.

HALO

This almost spherical region contains scattered stars, star clusters, and unseen dark matter.

MILKY WAY

UNIVERSAL ARCHITECTURE

Galaxies are collections of millions to hundreds of billions of stars linked together by mutual gravity and the gravity of dark matter. They may occur in galaxy clusters – our Milky Way, for example, is a member of the Local Group, a sparse cluster with about 50 members. Denser clusters, meanwhile, can contain thousands of closely packed galaxies surrounded by hot gas and abundant dark matter. At the largest known scale of organization, superclusters (clusters of clusters) may form walls of galaxies that extend hundreds of millions of light-years across, but have a depth of only tens of millions of light-years.

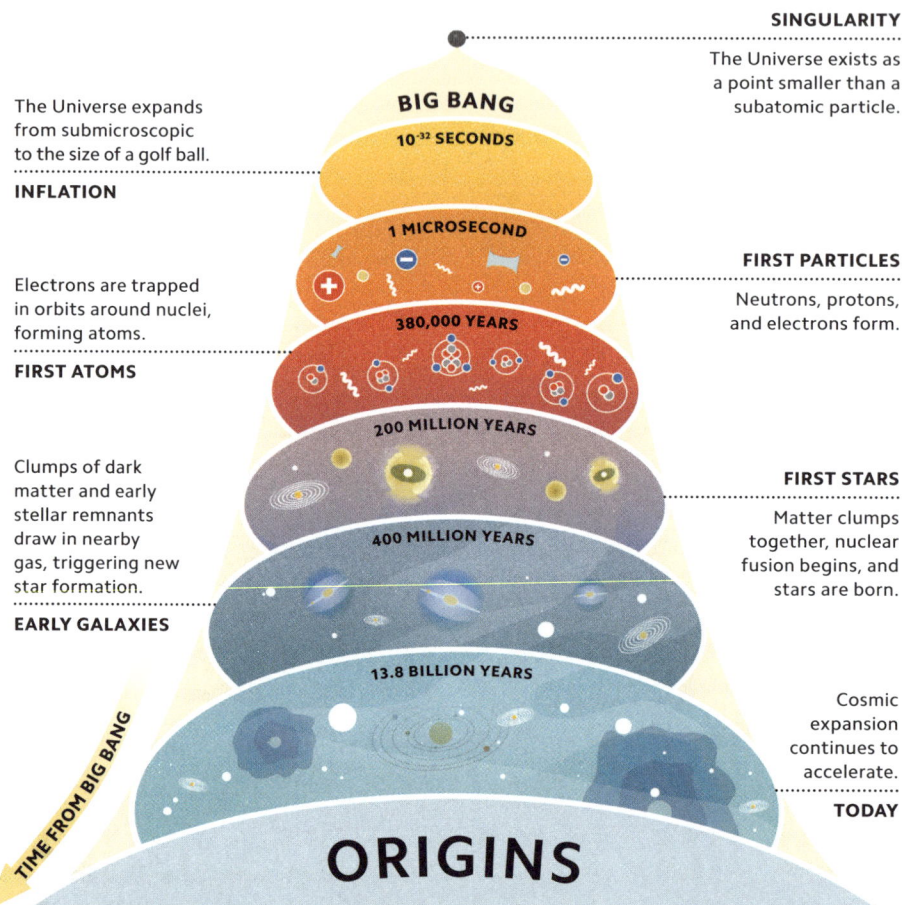

SINGULARITY

The Universe exists as a point smaller than a subatomic particle.

BIG BANG

10⁻³² SECONDS

The Universe expands from submicroscopic to the size of a golf ball.

INFLATION

1 MICROSECOND

FIRST PARTICLES

Neutrons, protons, and electrons form.

Electrons are trapped in orbits around nuclei, forming atoms.

FIRST ATOMS

380,000 YEARS

200 MILLION YEARS

FIRST STARS

Matter clumps together, nuclear fusion begins, and stars are born.

Clumps of dark matter and early stellar remnants draw in nearby gas, triggering new star formation.

400 MILLION YEARS

EARLY GALAXIES

13.8 BILLION YEARS

Cosmic expansion continues to accelerate.

TODAY

TIME FROM BIG BANG

ORIGINS

The Universe is expanding at ever greater speed, propelled by dark energy (see p.147) – a force that repels gravity. This and many observations support a theory for the origin of the Universe known as the Big Bang. It proposes that around 13.8 billion years ago, the entire Universe was compressed to a hot, dense state smaller than an atom, within which matter and energy were indistinguishable. As space expanded, rapid cooling allowed matter to become stable, and particles were eventually able to form atoms. Visible matter coalesced to form the first stars and galaxies.

Some matter attracted
by the black hole is fired
back out into space at
near-light speed.

PARTICLE JET

MATTER JOINING DISC

ACCRETION DISC

The black hole pulls
matter towards it; this
matter forms a bright and
hot spinning disc. It is
eventually pulled beyond
the event horizon.

BLACK HOLE

EVENT HORIZON

This is effectively the
boundary of the black hole
beyond which gravity is
so strong that light and
matter cannot escape.

At the core of a black hole,
matter is – theoretically –
crushed into infinite density.

SINGULARITY

POINTS OF ATTRACTION

A black hole is an object so dense that – at a certain range – nothing,
not even light, can escape its gravitational pull. Most black holes form
when the collapsing core of a monster star (see p.151) has so much mass
that, during a supernova explosion, gravity pulls it into a tiny point
called a singularity. Stellar black holes are relatively common, with
thousands in the Milky Way alone, but supermassive black holes
are rarer. These objects are found in the centres of many galaxies,
where they formed alongside the galaxies themselves, feeding
on gas, dust, and stars to grow to enormous size.

INDEX

Page numbers in **bold** refer to main entries.

diffusion **104**
diodes 67, 68
direct current 64
diseases 90, 123
divergent boundaries 130
diversity 89
DNA (deoxyribonucleic acid) 89, 101, **106**
  chromosomes 108
  epigenetics 112
  genes 109, 110
  genetic mutation 121
  meiosis 116–17
  mitosis 114–15
drugs 35

E

Earth 16, **126–43**, 148
earthquakes 130, 131, **133**
ecosystems **124**, 125
eggs 116, 119
Einstein, Albert 80, 81, 84
electric current 40, **64**
  circuits **66**
  semiconductors **67**
  transistors **68**
electrical energy 40, 41, 42
electricity, static 41, **65**
electromagnetism **63**
  electromagnetic interaction **72**
  electromagnetic radiation 45, 150
  electromagnetic spectrum **70–71**
  electromagnetic waves 40, 55, 58, 59, 63
electrons 9, 12, **13**, 17, 62, 76, 154
  circuits 66
  covalent bonds 19
  electric current 64
  electromagnetism 63
  emission spectra 77
  ionic bonds 18, 19
  metallic bonds 20

  movement of 40
  in orbitals 79
  static electricity 65
  wave functions 78
electrostatic attraction 18, 20
elementary particles **74–75**
elements 12, **16**, 18, 19, 26, 82
  periodic table **14–15**, 17
emission spectra 77
endoplasmic reticulum 92, 94, 96
endothermic reactions 28
energy 10, 44, 76, 77, 84, 98
  conservation of **42**
  quantum mechanics 76–79
  thermodynamics 42, **46–47**
  types of **40–41**
  and work **39**, 41
entropy 46, 47
enzymes 30, **99**, 110, 113
epigenetics **112**
equilibrium, chemical **27**
eukaryotes **89**, 92, 93, 95, 96, 97, 108
exosphere 136
exothermic reactions 28, 31
explosions 28, **31**
extinction **123**
eye colour 118

F

fault boundaries 130, 133
fermions **74**, 75
Ferrel cells 138
fertilization 116, 119
fission, nuclear **85**
fluids **56**
food web **125**
forces **38**, 48, 49, **72–73**, 75
  and circular motion 54
  magnetism **62**
  and motion 52
  work 39
fossil fuels 140, 142

frequency 58, 71
fundamental interactions (forces) 72–73
fungi **96**, 125
fusion, nuclear **84**, 151, 152

G

galaxies 146, 147, **153**, 154, 155
gametes 116, 117, 119
gamma rays 83
gases 10, 11, 22, 23, 25, 40, 45, 104
  gas laws 56
gauge bosons 72, 75
general theory of relativity **81**
genes **109**, 119, 120
  epigenetics **112**
  gene expression **110–11**
  genetic code 106
  genetic engineering 113
  genetic mutation 121
  genomes **120**
  inheritance **118**
  meiosis **116–17**
  mitosis **114–15**
genomes **120**
genotype 118, 120
global warming **142**, 143
gravity 38, 41, **48**, 54, 55, 72, **73**, 81
  in the Universe 150, 152, 153, 154, 155
greenhouse effect **141**
greenhouse gases 141, 142
guanine (G) 106, 107, 109

H

habitats **124**
Hadley cells 138
hadrons 74
half-lives 82
haploid 117, 119

ACKNOWLEDGMENTS

DK would like to thank the following for
their help with this book: Katie John for
proofreading; Vanessa Bird for the index;
and Sarah Binns for her advice on the
contents list. Senior DTP Designer: Harish
Aggarwal. Senior jackets coordinator:
Priyanka Sharma Saddi.

All images © Dorling Kindersley
For further information see:
www.dkimages.com